Synchronization in Complex Networks of Nonlinear Dynamical Systems

Synchronization in Complex Networks *of* Nonlinear Dynamical Systems

Chai Wah WU

IBM Thomas J. Watson Research Centre, USA

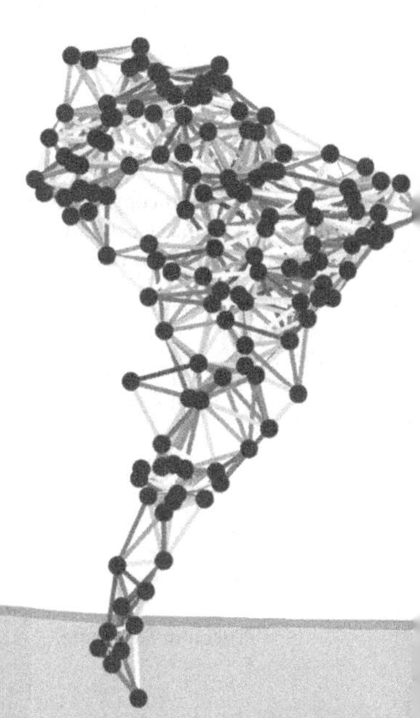

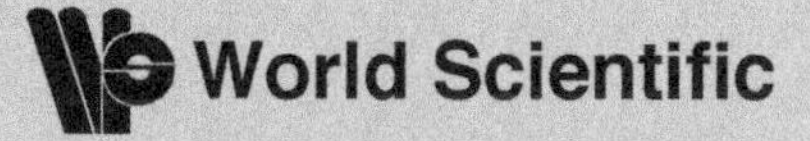

World Scientific

NEW JERSEY • LONDON • SINGAPORE • BEIJING • SHANGHAI • HONG KONG • TAIPEI • CHENNAI

Published by

World Scientific Publishing Co. Pte. Ltd.
5 Toh Tuck Link, Singapore 596224
USA office: 27 Warren Street, Suite 401-402, Hackensack, NJ 07601
UK office: 57 Shelton Street, Covent Garden, London WC2H 9HE

British Library Cataloguing-in-Publication Data
A catalogue record for this book is available from the British Library.

SYNCHRONIZATION IN COMPLEX NETWORKS OF NONLINEAR DYNAMICAL SYSTEMS

ISBN-13 978-981-270-973-8
ISBN-10 981-270-973-8

Printed in Singapore.

To Ann and Brian

For keeping my chaotic life in synch

Preface

This book studies complete synchronization of coupled nonlinear systems in arbitrary networks and continues where the book "Synchronization in coupled chaotic circuits and systems" (World Scientific, 2002) left off. In particular, we delve more deeply into the connection between coupling topology and synchronization and focus on the graph-theoretical properties of the underlying topology. Another aspect of this book is to study how properties of complex network models such as small world models or preferential attachment models affect the synchronization properties of a network of dynamical systems coupled via such graphs. This area of research is experiencing tremendous growth of activity and no doubt many new results will have appeared by the time you read this and we apologize beforehand for this book's incompleteness.

This book would not have been possible without the discussions, advice and encouragement of many friends and colleagues. I would especially like to thank Ming Cao, Guanrong Chen, Leon Chua, Don Coppersmith, Alan Hoffman, Ying Huang, Ljupco Kocarev, Tamás Roska, Mike Shub, and Charles Tresser for stimulating intellectual interactions over the years.

I would also like to thank International Business Machines Corporation for giving me the freedom and support to work on this subject over the last dozen years. Last but not least, I would like to thank Ann and Brian for their patience, love and support over the years.

May 2007
Yorktown Heights, New York
Chai Wah Wu

Contents

Chapter 1

Introduction

In the last two decades or so, there has been a resurgence in the analysis of the behavior in complex networks of interacting systems. During this time, two parallel branches of research activities have emerged. On the one hand, starting with the 1983 paper by Fujisaka and Yamada on synchronization in coupled systems [Fujisaka and Yamada (1983)] and subsequently the 1990 paper by Pecora and Carroll on synchronization in chaotic systems [Pecora and Carroll (1990)], there has been a plethora of activity on the synchronization of coupled chaotic systems. In this form of synchronization, called *complete* synchronization[1], the state variables of individual systems converge towards each other. Complete synchronization is more restrictive than phase synchronization that was studied as early as the 17th century by Huygens [Bennett *et al.* (2002)], and can be easier to analyze. In the last few years, agreement and consensus protocols of interconnected autonomous agents have been actively studied in the control systems community. In these problems the goal is to have the states of the agents agree to each other. For instance, the state can be the heading of a mobile agent in flocking problems where the goal is to have all agents move in the same direction. These problems, whose state equations form a linear system, can also be considered as a synchronization problem.

On the other hand, novel models of random graphs have been proposed to study the complex networks that we observe around us. In 1998 Watts and Strogatz proposed a model of a small-world network [Watts and Strogatz (1998)] and in 1999 Albert and Barabási proposed a model of a scale-free network based on preferential attachment [Barabási and Albert (1999)]. These graph models mimic complex networks in natural and man-made systems more accurately than the classical random graph mod-

[1]This is sometimes also referred to as identical synchronization

els studied by Rapoport [Rapoport (1957)] and by Erdös and Renyi [Erdös and Renyi (1959)] in the late 1950's. Examples of such networks include communication networks, transportation networks, neural networks and social interaction networks. Although features of these networks have been studied in the past, see for example Milgram's letter passing experiments [Milgram (1967)] and Price's citation network model [de Solla Price (1965, 1976)], it was only recently that massive amount of available data and computer processing power allow us to more easily analyze these networks in great detail and verify the applicability of various models.

The present book studies the intersection of these two very active research areas. In particular, the main object of study is synchronization phenomena in networks of coupled dynamical systems where the coupling topology can be expressed as a complex network. We attempt to combine recent results in these two interdisciplinary areas to obtain a view of such synchronization phenomena.

The focus of this book is on complex interacting systems that can be modelled as an interconnected network of identical systems. In particular, we are interested in the relationship between the coupling topology and the ability to achieve coherent behavior in the network. A typical interconnected system is illustrated in Fig. 1.1. We characterize the coupling topology by means of a directed graph, called the *interaction graph* . If system i influences system j, then there is a directed edge (i, j) starting from system i and ending in system j. In this case, the circles in Fig. 1.1 are vertices and the arrows are edges of this graph. We consider weighted directed graphs, where the weight of an edge indicates the coupling strength of that connection. By definition, the interaction graph of a network is *simple*, i.e. it does not contain self-loops from a vertex to itself and there is at most one edge between vertices.

One of the main results in this text is the following intuitive conclusion. If there is a system (called the *root system*) which influences directly or indirectly all other systems, then coherent behavior is possible for sufficiently strong coupling. The root system can change with time and the ability of the root system to influence all other systems can occur at each moment in time, or through a number of time steps. To reach this conclusion we use tools from dynamical systems, graph theory, and linear algebra.

This text is organized as follows. Chapter 2 summarizes basic graph theory, properties of graphs and linear algebra that we need. Detailed proofs of some of the results in Chapter 2 are found in Appendix A.

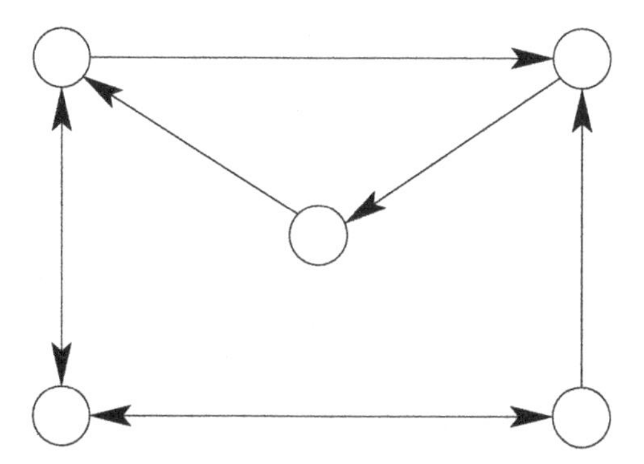

Fig. 1.1 Network of interconnected systems. Circles indicate individual systems and arrows indicate the coupling between them.

In Chapter 3 we study various models of graphs that have been proposed to model man-made and naturally occurring networks.[2]

In Chapter 4 we study synchronization in a network of nonlinear continuous time dynamical systems. The case of discrete-time systems will be addressed in Chapter 5. In these 2 chapters we establish a relationship between synchronizability and properties of the interaction graph.

In Chapter 6 we consider the special case where the coupling and the individual systems are linear. In particular, we show how the discrete time case is related to ergodicity of inhomogeneous Markov chains.

Finally, in Chapter 7 we study several consensus and agreement problems of autonomous agents that have been formulated as a network of coupled nonautonomous linear systems.

[2]We use the words *graph* and *network* interchangeably.

Chapter 2

Graphs, Networks, Laplacian Matrices and Algebraic Connectivity

In this chapter we summarize some definitions and results from graph theory. In particular, we focus on results about the Laplacian matrix and algebraic connectivity of graphs which are of great utility in deriving synchronization properties of networks of coupled dynamical systems.

2.1 Graphs and digraphs

A graph $\mathcal{G} = (V, E)$ consists of a set of vertices V and a set of edges E. Each edge is a subset of V of two elements denoting a connection between the two vertices, i.e. $E \subset \{\{i,j\} : i,j \in V\}$. If the two vertices in an edge are the same, we call this edge a *self-loop*. We restrict ourselves to finite graphs, i.e. graphs where V and E are finite sets. In general, we assume $V = \{1, \ldots, n\}$, where n is the number of vertices of the graph, also called the *order* of the graph. A *simple* graph is a graph with no self-loops and no multiple edges between the same pair of vertices. A simple graph with an edge between any pair of distinct vertices is a complete graph. A clique of a graph is defined as a subgraph which is complete (Fig. 2.1). The sum $\mathcal{G} + \mathcal{H}$ of the graphs $\mathcal{G}$ and $\mathcal{H}$ is the graph whose adjacency matrix is the sum of the adjacency matrices of $\mathcal{G}$ and $\mathcal{H}$. For two graphs $\mathcal{G}_1 = (V, E_1)$ and $\mathcal{G}_2 = (V, E_2)$ with the same vertex set, the union is defined as $\mathcal{G}_1 \cup \mathcal{G}_2 = (V, E_1 \cup E_2)$. Let $\mathcal{G} = (V_\mathcal{G}, E_\mathcal{G})$ and $\mathcal{H} = (V_\mathcal{H}, E_\mathcal{H})$. The Cartesian product $\mathcal{G} \times \mathcal{H}$ of the graphs $\mathcal{G}$ and $\mathcal{H}$ is a graph with vertex set $V_\mathcal{G} \times V_\mathcal{H}$ such that (a, b) is adjacent to (c, d) in $\mathcal{G} \times \mathcal{H}$ if and only if at least one of the following conditions are satisfied:

- $a = c$ and b is adjacent to d in $\mathcal{H}$;
- $b = d$ and a is adjacent to c in $\mathcal{G}$.

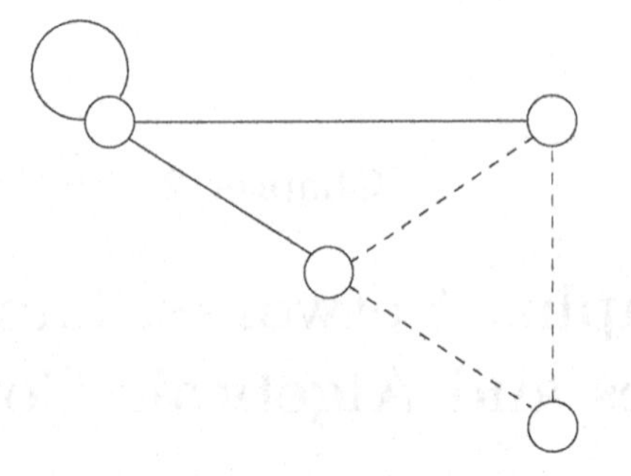

Fig. 2.1 A graph with a clique denoted by the dashed edges.

A *directed* graph (or digraph) is a graph where the edges are directed, i.e. each edge is an *ordered* pair of vertices with (i, j) denoting an edge which starts at vertex i and ends at vertex j. A mixed graph is a graph with both undirected edges and directed edges. In general, graphs are weighted, i.e. a positive weight is associated to each edge. An undirected or mixed graph can be considered as a directed graph by associating each undirected edge with weight w between vertex i and j as two directed edges between i and j with opposite orientation and weight w (Fig. 2.2). To distinguish graphs from digraphs, we also refer to graphs as *undirected* graphs. The *reversal* of a digraph is the digraph obtained by reversing the orientation of all the edges.

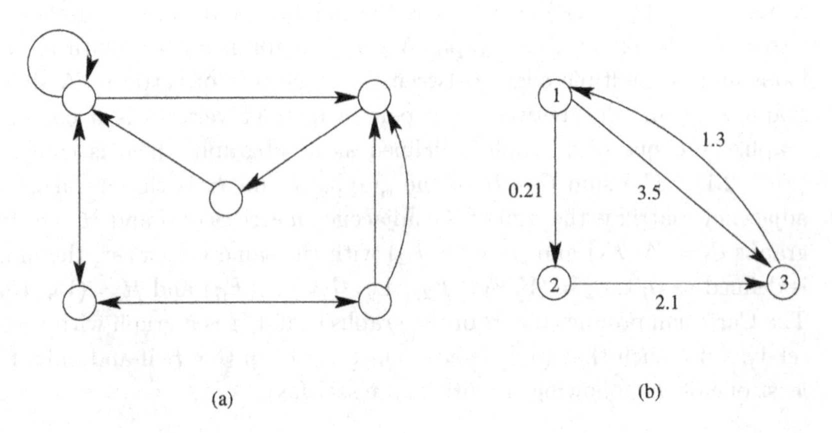

Fig. 2.2 Example of graphs. An undirected edge is denoted with arrowheads at both ends. (a) a non-simple mixed graph. (b) a weighted directed graph.

A graph is connected if there is a path between any pair of vertices. A digraph is *weakly connected* if ignoring the orientation of the edges, the resulting undirected graph is connected. A digraph is *strongly connected* if for any pair of vertices (i, j), there is a directed path from vertex i to vertex j. A digraph is *quasi-strongly connected* if for any pair of vertices (i, j), there exists a vertex k (which could be equal to i or j) with a directed path from vertex k to vertex i and a directed path from vertex k to vertex j. The distance between vertex a and vertex b is the sum of the weights of a directed path from a to b or from b to a minimized over all such paths. The diameter of a graph is the distance between two vertices, maximized over all pairs of vertices.

A directed tree is a digraph with n vertices and $n - 1$ edges with a root vertex such that there is a directed path from the root vertex to every other vertex. A spanning directed tree of a graph is a subgraph which is a directed tree with the same vertex set.

In Figure 2.3, a digraph is shown with the edges of a spanning directed tree shown in dashed lines. In general, the spanning directed tree is not unique.

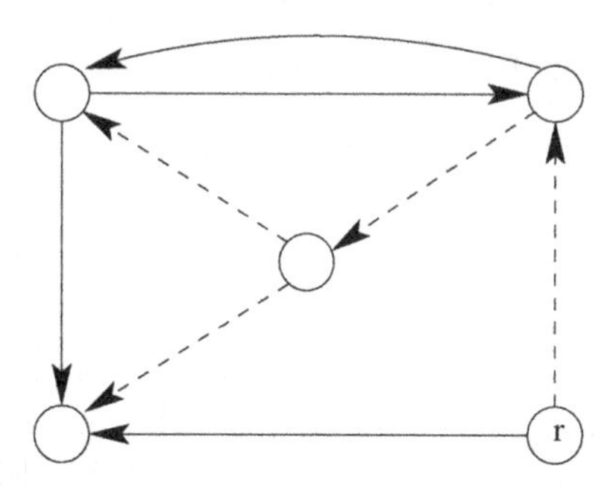

Fig. 2.3 Digraph with a spanning directed tree shown in dashed lines. The root of the tree is labeled **r**.

Theorem 2.1 *A digraph is quasi-strongly connected if and only if it contains a spanning directed tree.*

Proof: This is a well-known result, see e.g. [Swamy and Thulasiraman (1981)]. One direction in clear. As for the other direction, suppose a digraph is quasi-strongly connected. Starting with the list of vertices V, for every pair of distinct vertices a and b we replace them with a vertex $c(a, b)$ (which can be equal to a or to b) such that there are directed paths from $c(a, b)$ to a and to b. By the definition of quasi-strongly connectedness, we can repeat this reduction until there is only one vertex r left. By the construction, there are directed paths from r to every vertex in the graph. This implies that the graph contains a spanning directed tree. □

2.2 Matrices and graphs

There exists an intimate relationship between graph theory and matrix theory with both fields benefiting from insights in the other.

A (not necessarily symmetric) real matrix A is positive definite (semidefinite) if $x^T Ax > 0$ (≥ 0) for all nonzero real vectors x. We denote this as $A \succ 0$ ($A \succeq 0$). Negative definite matrices are defined similarly. A real matrix A is positive definite (semidefinite) if and only if all eigenvalues of $A + A^T$ are positive (nonnegative). The Kronecker product[1] of an n by m matrix A and a p by q matrix B is the np by mq matrix $A \otimes B$ defined as:

$$A \otimes B = \begin{pmatrix} A_{11}B & \cdots & A_{1m}B \\ \vdots & \ddots & \\ A_{n1}B & & A_{nm}B \end{pmatrix}$$

A matrix is reducible if it can be written as

$$P \begin{pmatrix} A & B \\ 0 & C \end{pmatrix} P^T$$

where P is a permutation matrix. A matrix is irreducible if it is not reducible. If A is irreducible, then for any $i \neq j$, there exists $i = i_1$, i_2, $\ldots$, $i_k = j$, such that $A_{i_h i_{h+1}} \neq 0$ for all h. Thus a digraph is strongly connected if and only if its adjacency matrix (or its Laplacian matrix) is irreducible. The graph of a nonnegative matrix A is defined as the weighted digraph with an edge with weight A_{ij} from vertex i to vertex j if and only if $A_{ij} \neq 0$.

Definition 2.2 The *interaction graph* of a matrix A is defined as the graph of A^T.

Thus the interaction graph and the graph of a matrix are reversals of each other. The definition of interaction graph is useful and makes more sense when considering networks of coupled dynamical systems with linear coupling. It coincides with the definition in Chapter 1 for the state equations we will consider in Chapters 4 and 5.

[1] Also called direct product or tensor direct product.

Conversely, the *adjacency* matrix A of a (di)graph is a nonnegative matrix defined as $A_{ij} = w$ if and only if (i, j) is an edge with weight w. The outdegree $d_o(v)$ of a vertex v is the sum of the weights of edges emanating from v. The indegree $d_i(v)$ of a vertex v is the sum of the weights of edges into v. A vertex is balanced if its outdegree is equal to its indegree. A graph is balanced if all its vertices are balanced. The maximum outdegree and indegree of a graph are denoted as Δ_o and Δ_i respectively. The minimum outdegree and indegree of a graph are denoted as δ_o and δ_i respectively.

The *Laplacian matrix* of a graph is a zero row sums nonnegative matrix L defined as $L = D - A$, where A is the adjacency matrix and D is the diagonal matrix of vertex outdegrees[2]. For instance, the adjacency matrix and the Laplacian matrix of the graph in Fig. 2.2b are given by:

$$A = \begin{pmatrix} 0 & 0.21 & 3.5 \\ 0 & 0 & 2.1 \\ 1.3 & 0 & 0 \end{pmatrix}, \quad L = \begin{pmatrix} 3.71 & -0.21 & -3.5 \\ 0 & 2.1 & -2.1 \\ -1.3 & 0 & 1.3 \end{pmatrix}$$

It is easy to see that a graph is balanced if and only if its Laplacian matrix has zero row sums and zero column sums. For a (undirected) graph of n vertices and m edges, the *incidence* matrix is defined as the n by m $0-1$ matrix M where $M_{ij} = 1$ if and only if vertex i is incident to the j-th edge. For a directed graph, the incidence matrix is defined as

$$\begin{aligned} M_{ij} &= -1 \text{ if the } j\text{-th edge starts at vertex } i \\ M_{ij} &= 1 \quad \text{if the } j\text{-th edge ends at vertex } i \\ M_{ij} &= 0 \quad \text{otherwise} \end{aligned}$$

Without loss of generality, we assume that $A_{ij} \leq 1$. An exception to this assumption is when a graph is *unweighted*, defined as the case when A_{ij} are natural numbers, with $A_{ij} = k$ denoting k edges from vertex i to vertex j. If $A_{jk} \neq 0$, then vertex j is the *parent* of vertex k and vertex k is the *child* of vertex j. The complement of a graph $\mathcal{G}$ without multiple edges is defined as the graph $\overline{\mathcal{G}}$ with the same vertex set as $\mathcal{G}$ and adjacency matrix $\overline{A}$ where $\overline{A}_{ij} = 1 - A_{ij}$ for $i \neq j$.

It can be shown that the Laplacian matrix of a graph can be written as $L = MM^T$ where M is the incidence matrix of the digraph obtained by assigning an arbitrary orientation to each edge (see e.g. the proof of Lemma 2.17).

[2]There are other ways to define the Laplacian matrix of a directed graph, see for instance [Bapat *et al.* (1999)].

Definition 2.3 The Frobenius normal form of a square matrix L is:

$$L = P \begin{pmatrix} B_1 & B_{12} & \cdots & B_{1k} \\ & B_2 & \cdots & B_{2k} \\ & & \ddots & \vdots \\ & & & B_k \end{pmatrix} P^T \tag{2.1}$$

where P is a permutation matrix and B_i are square irreducible matrices.

Lemma 2.4 *The matrices B_i are uniquely determined by L although their ordering can be arbitrary as long as they follow a partial order induced by $\lhd$ which is defined as $B_i \lhd B_j \Leftrightarrow B_{ij} \neq 0$.*

Proof: This is due to the fact that the matrices B_i correspond to strongly connected components of the graph of L which are unique. See [Brualdi and Ryser (1991)] for a proof. □

We define the degree of reducibility of L in Eq. (2.1) as $k - 1$. The decomposition of a matrix into Frobenius normal form is equivalent to the decomposition of the corresponding graph into strongly connected components and this can be done in linear time using standard search algorithms [Tarjan (1972)]. The partial order in Lemma 2.4 defines a condensation directed graph.

Definition 2.5 [Brualdi and Ryser (1991)] The condensation directed graph of a directed graph $\mathcal{G}$ is constructed by assigning a vertex to each strongly connected component of $\mathcal{G}$ and an edge between 2 vertices if and only if there exists an edge of the same orientation between corresponding strongly connected components of $\mathcal{G}$.

A directed graph and its corresponding condensation digraph is shown in Fig. 2.4.

Lemma 2.6 *The condensation directed graph $\mathcal{H}$ of $\mathcal{G}$ contains a spanning directed tree if and only if $\mathcal{G}$ contains a spanning directed tree.*

Proof: Suppose that $\mathcal{H}$ contains a spanning directed tree. For each edge (i, j) in $\mathcal{H}$, we pick an edge from the i-th strongly connected component to the j-th strongly connected component. We also pick a spanning directed tree inside each of the strongly connected component of $\mathcal{G}$ such that the edges above end at the roots of these trees. It is clear that such a choice is also possible. Adding these edges to the spanning directed trees we obtain a spanning directed tree for $\mathcal{G}$. If $\mathcal{H}$ does not contain a spanning directed tree, then there exists at least 2 vertices in $\mathcal{H}$ with indegree zero. This means

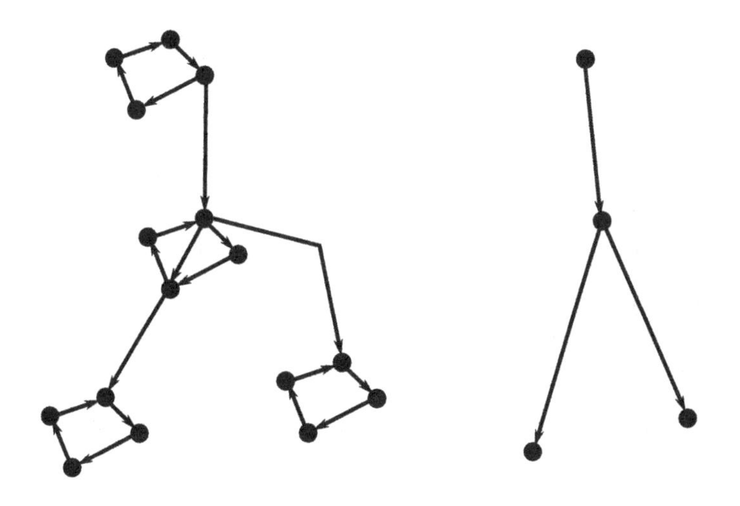

Fig. 2.4 A directed graph and the corresponding condensation directed graph.

that there are no edges into at least two strongly connected components and thus no spanning directed tree can exist in $\mathcal{G}$. □

The following result on nonnegative matrices forms the core of Perron-Frobenius theory [Minc (1988)] and it will be useful in analyzing Laplacian matrices.

Theorem 2.7 *A nonnegative matrix has a real eigenvalue $r \geq 0$ such that $r \geq |\lambda_i|$ for all eigenvalues λ_i and there exists an eigenvector corresponding to r which has only nonnegative components. If in addition the matrix is irreducible, then $r > 0$ and is a simple eigenvalue and there exists a corresponding eigenvector with only positive components.*

The following results are useful in bounding the eigenvalues of the Laplacian matrix. First is the classical Lévy-Desplanques nonsingularity criterion of matrices [Lévy (1881); Desplanques (1887)]:

Theorem 2.8 *If A is square matrix such that $|A_{ii}| > \sum_{j \neq i} |A_{ij}|$ for all i, then A is nonsingular.*

Theorem 2.8 is equivalent to the Gershgorin circle criterion [Geršgorin (1931)] for localizing eigenvalues:

Theorem 2.9 *All eigenvalues of a square matrix A lie in the region:*

$$\bigcup_i \left\{ x : |A_{ii} - x| \leq \sum_{j \neq i} |A_{ij}| \right\}$$

The following theorem in [Taussky (1949)] provides a stronger criterion than Theorem 2.8 in the case the matrix is irreducible:

Theorem 2.10 *If A is an irreducible square matrix such that $|A_{ii}| \geq \sum_{j \neq i} |A_{ij}|$ for all i with the inequality strict for at least one i, then A is nonsingular.*

The following result bounds the eigenvalues of stochastic matrices[3].

Theorem 2.11 *For an eigenvalue λ of an n by n stochastic matrix, the argument ϕ of $1 - \lambda$ satisfies*

$$\frac{\pi}{n} - \frac{\pi}{2} \leq \phi \leq \frac{\pi}{2} - \frac{\pi}{n}$$

Proof: This is a consequence of the results in [Dmitriev and Dynkin (1945)][4]. For instance, this result follows from Theorem 1.7 in [Minc (1988), page 175]. □

Theorem 2.11 is illustrated schematically in Figure 2.5. All eigenvalues of an n by n stochastic matrix are located in the shaded region.

Corollary 2.12 *The argument ϕ of all eigenvalues of a n by n Laplacian matrix L satisfies*

$$\frac{\pi}{n} - \frac{\pi}{2} \leq \phi \leq \frac{\pi}{2} - \frac{\pi}{n}$$

Proof: Follows from Theorem 2.11 and the fact that $I - \alpha L$ is stochastic for a small enough $\alpha > 0$. □

A substochastic matrix is a nonnegative matrix whose rows sum to less than or equal to 1. Let n_F denote the maximal order of the diagonal blocks B_k in the Frobenius normal form (Definition 2.3), i.e. n_F is the size of the largest strongly connected component of the corresponding graph.

[3] Throughout this text stochastic matrices refer to row-stochastic matrices, i.e. nonnegative matrices where the entries in each row sum to 1.

[4] The results in [Dmitriev and Dynkin (1945)] were generalized in [Karpelevich (1951)] to completely characterize the region in the complex plane where eigenvalues of stochastic matrices are located. See also [Kellogg (1972)] for a sharpening of this bound when the graph of the matrix does not contain a directed Hamiltonian circuit (i.e. a graph like Fig. 2.9(f)).

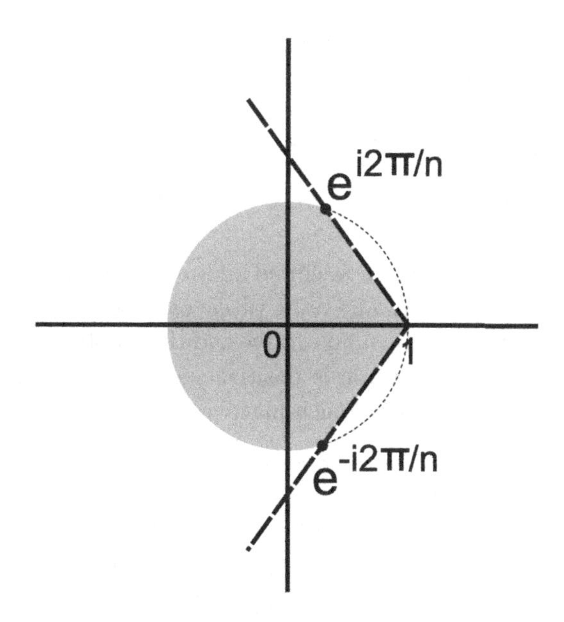

Fig. 2.5 Localization of eigenvalues of a stochastic matrix. All eigenvalues of a n by n stochastic matrix are located inside the shaded region.

Theorem 2.13 *For an eigenvalue λ of a stochastic matrix, the argument ϕ of $1 - \lambda$ satisfies*

$$\frac{\pi}{n_F} - \frac{\pi}{2} \leq \phi \leq \frac{\pi}{2} - \frac{\pi}{n_F}$$

Proof: First note that the proof of Theorem 1.7 in [Minc (1988), page 175] is also valid for substochastic matrices. Next note that the eigenvalues of a stochastic matrix are the eigenvalues of the diagonal blocks B_k in the Frobenius normal form which are substochastic matrices. □

Corollary 2.14 *The argument ϕ of all eigenvalues of a Laplacian matrix L satisfies*

$$\frac{\pi}{n_F} - \frac{\pi}{2} \leq \phi \leq \frac{\pi}{2} - \frac{\pi}{n_F}$$

A consequence of Theorem 2.13 and Corollary 2.14 is that the eigenvalues of a stochastic matrix or a Laplacian matrix are real if the strongly connected components of the corresponding graph have at most 2 vertices each.

Definition 2.15 Let M_1 be the class of real zero row sums matrices where each row has exactly two nonzero entries. Let M_2 be the class of matrices $A \in M_1$ such that for any pair of indices i and j there exist indices $i_1, i_2, \ldots, i_l$ with $i_1 = i$ and $i_l = j$ such that for all $1 \leq q < l$, there exists p such that $A_{p,i_q} \neq 0$ and $A_{p,i_{q+1}} \neq 0$.

Definition 2.16 The vector $\mathbf{1}$ is defined as the vector of all 1's, i.e. $\mathbf{1} = (1, \ldots, 1)^T$.

The Laplacian matrix of a weighted undirected graph is a real symmetric matrix with zero row sums and nonpositive off-diagonal elements. By Theorem 2.9 all eigenvalues are nonnegative and this implies that the Laplacian matrix of an undirected graph is positive semidefinite. The following two results list some properties of such matrices.

Lemma 2.17 *A matrix A is a real symmetric zero row sums matrix with nonpositive off-diagonal elements if and only if $A = C^T C$ for some $C \in M_1$. A matrix A is a real symmetric irreducible zero row sums matrix with nonpositive off-diagonal elements if and only if $A = C^T C$ for some $C \in M_2$.*

Proof: The $(i,j)^{th}$ element of $C^T C$ is the inner product of the i-th and j-th columns of C. Let $C \in M_1$. If $i \neq j$, the terms in the inner product of the i-th and j-th columns of C is 0 or negative. Thus $C^T C$ has nonpositive off-diagonal elements. Since $C\mathbf{1} = 0$, $C^T C\mathbf{1} = 0$ and thus $C^T C$ is a symmetric zero row sums matrix.

Let A be a symmetric matrix with zero row sums and nonpositive off-diagonal elements. Construct C as follows. For each nonzero row of A we generate several rows of C of the same length as follows: for the i-th row of A, and for each $i < j$ such that $A_{ij} \neq 0$, we add a row to C with the i-th element being $\sqrt{-A_{ij}}$, and the j-th element being $-\sqrt{-A_{ij}}$. We claim that this matrix C satisfies $A = C^T C$. First note that since $C\mathbf{1} = 0$, $C^T C$ is a symmetric zero row sum matrix. Certainly $C \in M_1$. A_{ij} is the inner product between the i-th column and the j-column of C since by construction, there is only one row of C with nonzero entries in both the i-th and j-th position, giving the appropriate result. From the construction of C, it is easy to show that A is irreducible if and only if $C \in M_2$. □

Lemma 2.18 *If A is a symmetric zero row sums matrix with nonpositive off-diagonal elements, then $x^T A y = -\sum_{i<j} A_{ij}(x_i - x_j)(y_i - y_j)$. If in addition A is irreducible, then $x^T A x = 0$ if and only if $x_i = x_j$ for all i, j.*

Proof: By Lemma 2.17 $x^T A y$ can be written as $x^T C^T C y$. In particular, for each $i < j$ such that $A_{ij} \neq 0$ there corresponds an k-th element in Cy

of the form $\sqrt{-A_{ij}}(y_i - y_j)$. Similarly, the k-th element in Cx is of the form $\sqrt{-A_{ij}}(x_i - x_j)$. If A is irreducible, then $A \succeq 0$ and 0 is a simple eigenvalue by Theorem 2.7 and thus $x^T A x = 0$ implies x is in the kernel of A, i.e. $x = \alpha \mathbf{1}$. □

Definition 2.19 Consider a reducible matrix A of order n. The matrix A is n-reducible if it is diagonal. For $1 \leq m < n$, the matrix A is m-reducible if it is not $(m+1)$-reducible and it can be written as

$$A = P \begin{pmatrix} B_1 & B_{12} & \cdots & & & & B_{1,k+m} \\ & \ddots & & & & & \\ & & B_k & B_{k,k+1} & \cdots & & B_{k,k+m} \\ & & & B_{k+1} & 0 & & 0 \\ & & & & \ddots & & 0 \\ & & & & & & B_{k+m} \end{pmatrix} P^T \tag{2.2}$$

where P is a permutation matrix and B_i are square irreducible matrices.

Equivalently, a matrix A is m-reducible if and only if it can be written as Eq. (2.2) and for $i \leq k$, there exists $j > i$ such that $B_{ij} \neq 0$. A reducible matrix of order n is m-reducible for a unique $1 \leq m \leq n$.

Theorem 2.20 *A matrix A of order n is m-reducible if and only if $\mu_k(A) = 0$, $\mu_l(A) > 0, 1 \leq k \leq m < l \leq n$, where $\mu_m(A)$ is defined as*

$$\mu_m(A) = \min_{\emptyset \neq T_1,\ldots,T_m \neq V, T_i \cap T_j = \emptyset} \sum_{i=1}^{m} \sum_{j \in T_i, k \notin T_i} |A_{jk}|$$

Proof: First note that $\mu_m(A) \geq 0$. If A is m-reducible, then writing A in the form of Eq. (2.2) shows that $\mu_k(A) = 0$ for all $k \leq m$. On the other hand, if $\mu_m(A) = 0$, then it can be written as Eq. (2.2) since the submatrix corresponding to $V \backslash (T_1 \cup \cdots \cup T_m)$ can be written in Frobenius normal form, i.e. block triangular with irreducible diagonal blocks. □

Just as $\mu_1(A)$ is a measure of irreducibility for irreducible matrices [Fiedler (1972)], $\mu_{m+1}(A)$ can be considered as a measure of m-reducibility for a m-reducible matrix.

Theorem 2.21 *Let A be a reducible Laplacian matrix. Then A is m-reducible if and only if the multiplicity of the zero eigenvalue is m.*

Proof: If A is m-reducible, then $B_{k+1}, \ldots, B_{k+m}$ are irreducible Laplacian matrices and thus each contains a simple zero eigenvalue. Since A is not $(m+1)$-reducible, for $1 \leq i \leq k$, there exists $j > i$ such that $B_{ij} \neq 0$ and thus one of the row sums of B_i is positive and B_i is nonsingular by Theorem 2.10. Thus the zero eigenvalue has multiplicity m. If A is not m-reducible, it is k-reducible for some $m \neq k$ and thus the zero eigenvalue has multiplicity k. □

Corollary 2.22 *Let A be a reducible stochastic matrix. Then A is m-reducible if and only if the multiplicity of the eigenvalue* 1 *is m.*

From Theorem 2.7, all eigenvalues of the Laplacian matrix L are nonnegative, with 0 being a simple eigenvalue if the graph is strongly connected. This result can be strengthened as follows.

Theorem 2.23 *The zero eigenvalue of a reducible Laplacian matrix of a digraph has multiplicity m if and only if m is the minimum number of directed trees which together span the reversal of the digraph.*

Proof: Let A be a m-reducible matrix as in Eq. (2.2) and consider the reversal of the graph of A. This graph has m groups of vertices corresponding to $B_{k+1}, \ldots B_{k+m}$ which have no edges pointing towards them. Therefore any spanning directed forest must have the root of some directed tree in each group. Thus the spanning forest has at least m trees. $B_{k+1}, \ldots, B_{k+m}$ correspond to strongly connected components and for $i \leq k$, $B_{ij} \neq 0$ for some $j > i$. These facts, together with a consideration of the condensation directed graph show that there exists a spanning forest with m trees. Conversely, if A is not m-reducible, it is m'-reducible for $m' \neq m$ and thus m' is the minimal number of directed trees. □

Corollary 2.24 *A matrix A is irreducible or* 1*-reducible if and only if the reversal of the digraph of A contains a spanning directed tree.*

Corollary 2.25 *The zero eigenvalue of the Laplacian matrix of a digraph is simple if and only if the reversal of the digraph contains a spanning directed tree.*

A consequence of Theorem 2.9 or Corollary 2.12 is that all nonzero eigenvalues of a Laplacian matrix has strictly positive real parts. The following result is well known.

Corollary 2.26 *The zero eigenvalue of the Laplacian matrix of a graph has multiplicity m if and only if m is the number of connected components in the graph.*

By Perron-Frobenius theory (Theorem 2.7), irreducibility is a sufficient condition for the eigenvalue 1 in a stochastic matrix to be isolated. The following Corollary shows that by including 1-reducibility, we obtain a sufficient and necessary condition.

Corollary 2.27 *The eigenvalue 1 in a stochastic matrix A is isolated if and only if A is irreducible or 1-reducible. If the diagonal entries of A are all positive, then all eigenvalues other than 1 must have norm strictly less than 1.*

Proof: The first part follows from Corollary 2.22. The second part follows by using Theorem 2.9 to localize the eigenvalues. □

We next describe the structure of the left eigenvector of a reducible Laplacian matrix L corresponding to the zero eigenvalue. From Theorem 2.21, if L is m-reducible, then the zero eigenvalue has multiplicity m. We now exhibit m independent left eigenvectors corresponding to the zero eigenvalue. Without loss of generality, we assume that L can be written as

$$A = \begin{pmatrix} B_1 & B_{12} & \cdots & & & & B_{1,k+m} \\ & \ddots & & & & & \\ & & B_k & B_{k,k+1} & \cdots & & B_{k,k+m} \\ & & & B_{k+1} & 0 & & 0 \\ & & & & \ddots & & 0 \\ & & & & & & B_{k+m} \end{pmatrix} \tag{2.3}$$

i.e. Eq. (2.2) with $P = I$. Now B_1 is irreducible, and one of $B_{12}, \ldots, B_{1k}$ is not equal to the zero matrix. Since L has zero row sums this means that $(B_1)_{ii} \geq \sum_{i \neq j} |(B_1)_{ij}|$. $B_{1l} \neq 0$ for some l means that for at least one i, $(B_1)_{ii} > \sum_{i \neq j} |(B_1)_{ij}|$. By Theorem 2.10, B_1 is nonsingular. A similar argument shows that $B_2, \cdots, B_k$ are all nonsingular. Let $w^T L = 0$ and let w_i, $i = 1, \cdots, k+m$ be the subvectors of w corresponding to the components B_i. Then $w_1^T B_1 = 0$ which implies $w_1 = 0$. By induction, we can show that $w_2, \cdots, w_k = 0$. On the other hand, by Theorem 2.7, there exists a positive vector $\tilde{w}_i$ such that $\tilde{w}_i^T B_i = 0$ for $i = k+1, \cdots, k+m$ since each B_i is an irreducible Laplacian matrix. Thus for each $j = k+1, \cdots, k+m$, the following vector w is a nonzero eigenvector for the zero eigenvalue: $w_i = 0$ for $i \neq j$ and $w_j = \tilde{w}_j$. This forms an independent set of m eigenvectors.

Lemma 2.28 *For a graph $\mathcal{G}$ with Laplacian matrix L, there exists a positive vector $w > 0$ such that $w^T L = 0$ if and only if $\mathcal{G}$ is a disjoint union of strongly connected graphs.*

Proof: If $\mathcal{G}$ is a disjoint union of strongly connected graphs, each such subgraph $\mathcal{H}_i$ has a positive vector w_i such that $w_i^T L(\mathcal{H}_i) = 0$. Concatenating these vectors will form a positive vector w such that $w^T L(\mathcal{G}) = 0$. If $\mathcal{G}$ is not a disjoint union of strongly connected subgraphs, then $L(\mathcal{G})$ written in Frobenius normal form (Eq. 2.1) satisfies

(1) B_1 is irreducible,
(2) one of $B_{12}, \ldots, B_{1k}$ is not equal to the zero matrix.

Without loss of generality, we assume $P = I$ and the discussion above shows that B_1 is nonsingular and so any vector w such that $w^T L = 0$ must be of the form $(0 \quad w_2)^T$. □

Lemma 2.29 *Consider a Laplacian matrix L whose interaction graph contains a spanning directed tree. Let w be a nonnegative vector such that $w^T L = 0$. Then $w_i = 0$ for all vertices i that do not have directed paths to all other vertices in the interaction graph and $w_i > 0$ otherwise.*

Proof: Follows from the discussion above for the special case $m = 1$. In this case B_{k+1} corresponds exactly to all vertices that have directed path to all other vertices, i.e. they are roots of a spanning directed tree in the interaction graph. □

2.3 Algebraic connectivity

The concept of algebraic connectivity introduced in [Fiedler (1973)] is a useful quantity to characterize connectivity in a graph. In Chapters 4-5 we will show that the algebraic connectivity is also useful in deriving synchronization conditions in a network of coupled dynamical systems. For a Hermitian matrix A of order n, let the eigenvalues of A be arranged as:

$$\lambda_1(A) \leq \lambda_2(A) \leq \cdots \leq \lambda_n(A)$$

We will also write $\lambda_1(A)$ and $\lambda_n(A)$ as $\lambda_{\min}(A)$ and $\lambda_{\max}(A)$ respectively.

2.3.1 *Undirected graphs*

This is the original concept introduced by Fiedler.

Definition 2.30 The algebraic connectivity of an undirected graph with Laplacian matrix L is defined as $\lambda_2(L)$, the second smallest eigenvalue of L.

Of related interest is the largest eigenvalue of L, denoted as $\lambda_n(L)$. Some properties of $\lambda_2(L)$ and $\lambda_n(L)$ are:

Theorem 2.31 *For an undirected graph with minimum vertex degree δ and maximum vertex degree Δ,*

(1) $\lambda_2 \geq 0$ with the inequality strict if and only if the graph is connected.
(2) $\lambda_2 \leq \frac{n}{n-1}\delta \leq \frac{n}{n-1}\Delta \leq \lambda_n$.
(3) If an unweighted graph is not complete, then $\lambda_2 \leq \delta$.
(4) Let L_K be the Laplacian matrix of the complete graph. If $\alpha > 0$ and $\alpha L_K - L$ is positive semidefinite, then $\lambda_2(L) \geq \frac{n}{\alpha}$.

Proof: By considering undirected graphs as balanced directed graphs, properties (1)-(2) follow from Corollary 2.37 in Section 2.3.3. A proof of property (3) can be found in [Fiedler (1973)]. Note that $\lambda_2 \leq \delta$ is not true for a complete graph, since $\lambda_2 = n$ and $\delta = n - 1$. As for property (4) we follow the proof in [Guattery *et al.* (1997)]. Assume that $\alpha L - L_K \succeq 0$. Let v be the unit norm eigenvector of L corresponding to $\lambda_2(L)$. Since $L_K = nI - \mathbf{1}^T\mathbf{1}$, $x^T L_K x = nx^T x$ for all $x \perp \mathbf{1}$. This means that

$$0 \leq v^T(\alpha L - L_K)v = \alpha\lambda_2 - n$$

□

2.3.2 *Directed graphs*

Let $\mathbf{K}$ be the set $\{x \in \mathbb{R}^V, x\perp\mathbf{1}, \|x\| = 1\}$, i.e. the set of real vectors of unit norm in $\mathbf{1}^\perp$, the orthogonal complement of $\mathbf{1}$. Recall that a directed graph can be decomposed into strongly connected components corresponding to the Laplacian matrix L being written in Frobenius normal form (Eq. (2.1)). The square irreducible matrices B_i correspond to the strongly connected components of the graph.

Each square matrix B_i can be decomposed as $B_i = L_i + D_i$ where L_i is a zero row sums Laplacian matrix of the strongly connected components and D_i is diagonal. We can then write L as $L = L_s + L_r$ where L_s is the block diagonal matrix with L_i as diagonal blocks. In other words L_s and L_r are the Laplacian matrices of subgraphs $\mathcal{G}_s$ and $\mathcal{G}_r$ where $\mathcal{G}_s$ is a disjoint union of the strongly connected components and $\mathcal{G}_r$ is the acyclic residual subgraph whose edges are those edges in $\mathcal{G}$ which goes from one strongly connected component to another strongly connected component.

Let w_i be the unique positive vector such that $\|w_i\|_\infty = 1$ and $w_i^T L_i = 0$. The vector w_i exists by Theorem 2.7. Let W_i be the diagonal matrix with w_i on the diagonal. Let w be a nonnegative eigenvector such that $\|w\|_\infty = 1$ and $w^T L = 0$ and W be the diagonal matrix with w on the diagonal.

For digraphs, several generalizations of Fielder's algebraic connectivity have been proposed.

Definition 2.32 For a digraph with Laplacian matrix L expressed in Frobenius normal form, define[5]

- $a_1(L) = \min_{x \in \mathbf{K}} x^T L x = \min_{x \neq 0, x \perp \mathbf{1}} \frac{x^T L x}{x^T x}$
- $a_2(L) = \min_{x \in \mathbf{K}} x^T W L x$;
- If the graph is strongly connected, $a_3(L) = \min_{x \neq 0, x \perp \mathbf{1}} \frac{x^T W L x}{x^T \left(W - \frac{w w^T}{\|w\|_1}\right) x}$;
- $a_4(L) = \min_{1 \leq i \leq k} \eta_i$ where $\eta_i = \min_{x \neq 0} \frac{x^T W_i B_i x}{x^T W_i x}$ for $1 \leq i \leq k-1$ and $\eta_k = \min_{x \neq 0, x \perp \mathbf{1}} \frac{x^T W_k B_k x}{x^T \left(W_k - \frac{w_k w_k^T}{\|w_k\|_1}\right) x}$.

If $\mathcal{G}$ is a graph with Laplacian matrix L, we will also write $a_i(\mathcal{G})$ in place $a_i(L)$.

At first glance, $a_4(L)$ appears not well defined in Definition 2.32 because even though the matrices B_i are uniquely defined (up to simultaneous row and column permutation), their ordering within the Frobenius normal form (Eq. (2.1)) is not (Lemma 2.4). However, it is easy to see that the lower right block B_k is uniquely defined if and only if the reversal of the graph contains a spanning directed tree. In this case $a_4(L)$ is well-defined. If B_k is not uniquely defined, $a_4(L) = 0$ by Theorem 2.47 and $a_4(L) = 0$ for *any* admissible ordering of the B_i's within the Frobenius normal form. Thus $a_4(L)$ is well-defined for any Laplacian matrix L.

Similarly, λ_n can also be generalized to directed graphs:

Definition 2.33

- $b_1(L) = \max_{\|x\|=1} x^T L x$

Lemma 2.34 *For an undirected or balanced graph with Laplacian matrix* L, $a_1(L) = a_2(L) = a_3(L) = a_4(L) = \lambda_2(\frac{1}{2}(L + L^T))$, $b_1(L) = \lambda_{\max}(\frac{1}{2}(L + L^T))$.

Proof: Note that an undirected graph is balanced. The Laplacian matrix of a balanced graph have both zero row sums and zero column sums. This

[5]The vectors x in this section are all real vectors.

implies that $w_i = \mathbf{1}$ and $W = I$. Next note that $x^T Lx = \frac{1}{2}x^T(L+L^T)x$. Since $L + L^T$ is symmetric and $(L+L^T)\mathbf{1} = 0$, by the Courant-Fischer minmax theorem, $\lambda_2(L+L^T) = \min_{x\neq 0, x\perp \mathbf{1}} \frac{x^T(L+L^T)x}{x^Tx}$. To complete the proof, we note that for $x \perp \mathbf{1}$, $x^T\left(I - \frac{1}{n}\mathbf{1}\mathbf{1}^T\right)x = x^Tx$. □

2.3.3 *Basic properties of a_1 and b_1*

Some properties of the quantities a_i and b_1 are summarized in the following result.

Theorem 2.35

(1) a_1 can be efficiently computed as

$$a_1(L) = \min_{x\in\mathbb{R}^{n-1}, \|Qx\|=1} x^TQ^TLQx = \lambda_{\min}\left(\frac{1}{2}Q^T\left(L+L^T\right)Q\right)$$

where Q is an n by $n-1$ matrix whose columns form an orthonormal basis of $\mathbf{1}^\perp$;

(2) Let $\mathbf{T} = \{x \in \mathbb{R}^V, x \notin span(\mathbf{1})\}$ and let $L_K = nI - \mathbf{1}\mathbf{1}^T$ be the Laplacian matrix of the complete graph. If $\mathcal{G}$ is balanced,

$$a_1(\mathcal{G}) = n \min_{x\in\mathbf{T}} \frac{x^TLx}{x^TL_Kx} \leq n \max_{x\in\mathbf{T}} \frac{x^TLx}{x^TL_Kx} = b_1(\mathcal{G});$$

(3) If the reversal of the graph $\mathcal{G}$ does not contain a spanning directed tree, then $a_1(\mathcal{G}) \leq 0$;

(4) If the graph is not weakly connected, then $a_1(\mathcal{G}) \leq 0$;

(5) Let $\mathcal{G}$ be a balanced graph. Then $a_1(\mathcal{G}) > 0 \Leftrightarrow \mathcal{G}$ is connected $\Leftrightarrow \mathcal{G}$ is strongly connected;

(6) **Super- and sub-additivity.** *$a_1(\mathcal{G}+\mathcal{H}) \geq a_1(\mathcal{G}) + a_1(\mathcal{H})$ and $b_1(\mathcal{G}+\mathcal{H}) \leq b_1(\mathcal{G}) + b_1(\mathcal{H})$;*

(7) $a_1(\mathcal{G}\times\mathcal{H}) \leq \min(a_1(\mathcal{G}), a_1(\mathcal{H})) \leq \max(b_1(\mathcal{G}), b_1(\mathcal{H})) \leq b_1(\mathcal{G}\times\mathcal{H})$;

(8) $\lambda_1\left(\frac{1}{2}(L+L^T)\right) \leq a_1(L) \leq \lambda_2\left(\frac{1}{2}(L+L^T)\right)$, $\lambda_{n-1}\left(\frac{1}{2}(L+L^T)\right) \leq b_1(L) \leq \lambda_n\left(\frac{1}{2}(L+L^T)\right)$;

(9) If $\overline{L}$ is the Laplacian matrix of the complement $\overline{\mathcal{G}}$, then $a_1(L) + a_1(\overline{L}) = n$, where n is the order of L;

(10) If the off-diagonal elements of the adjacency matrix A of $\mathcal{G}$ are random variables chosen independently according to $P(A_{ij}=1) = p$, $P(A_{ij}=0) = 1-p$, then $a_1(\mathcal{G}) \approx pn$ in probability as $n \to \infty$.

Proof: See Theorem A.4 in Appendix A. □

The next result shows the relationship between a_1, b_1 and the degrees of vertices. Recall that $\Delta_o = \max_{v\in V} d_o(v)$, $\delta_o = \min_{v\in V} d_o(v)$, $\Delta_i = \max_{v\in V} d_i(v)$ and $\delta_i = \min_{v\in V} d_i(v)$.

Theorem 2.36 *Consider a graph $\mathcal{G}$ with adjacency matrix A and Laplacian matrix L.*

(1) Let v,w be two vertices which are not adjacent, i.e. $A_{vw} = A_{wv} = 0$. Then

$$a_1(L) \le \frac{1}{2}(d_o(v) + d_o(w)) \le b_1(L)$$

In particular, if the graph has two vertices with zero outdegrees, then $a(L) \le 0$.

(2)

$$\begin{aligned} a_1(L) &\le \min_{v\in V}\left\{d_o(v) + \frac{1}{n-1}d_i(v)\right\} \\ &\le \max_{v\in V}\left\{d_o(v) + \frac{1}{n-1}d_i(v)\right\} \le b_1(L) \end{aligned}$$

(3)

$$a_1(L) \le \min\left\{\delta_o + \frac{1}{n-1}\Delta_i, \Delta_o + \frac{1}{n-1}\delta_i\right\} \le \frac{n}{n-1}\min\{\Delta_o, \Delta_i\}$$

(4)

$$b_1(L) \ge \max\left\{\delta_o + \frac{1}{n-1}\Delta_i, \Delta_o + \frac{1}{n-1}\delta_i\right\} \ge \frac{n}{n-1}\max\{\delta_o, \delta_i\}$$

(5)

$$\frac{1}{2}\min_{v\in V}\{d_o(v) - d_i(v)\} \le a_1(L) \le \min_{v\ne w}\{d_o(v) + d_o(w)\}$$

(6)

$$b_1(L) \le \max_{v\in V}\left\{\frac{3}{2}d_o(v) + \frac{1}{2}d_i(v)\right\}$$

(7) Let $\mathcal{H}$ be constructed from a graph $\mathcal{G}$ by removing a subset of vertices with zero indegree from $\mathcal{G}$ and all adjacent edges. Then $a_1(\mathcal{H}) \ge a_1(\mathcal{G})$.

(8) Let $\mathcal{H}$ be constructed from $\mathcal{G}$ by removing k vertices from $\mathcal{G}$ and all adjacent edges. Then $a_1(\mathcal{H}) \geq a_1(\mathcal{G}) - k$.

(9) For a graph $\mathcal{G}$, let (V_1, V_2) be a partition of V and let $\mathcal{G}_i$ be the subgraph generated from V_i. Then

$$a_1(\mathcal{G}) \leq \min(a_1(\mathcal{G}_1) + |V_2|, a_1(\mathcal{G}_2) + |V_1|)$$

(10) If $\mathcal{G}$ is a directed tree and some vertex is the parent of at least two vertices, then $a_1(\mathcal{G}) \leq 0$. If the reversal of $\mathcal{G}$ is a directed tree then $a_1(\mathcal{G}) \leq \frac{d_i(r)}{n-1}$, where r is the root of the tree.

(11) A graph $\mathcal{G}$ with n vertices and m edges satisfies

$$a_1(\mathcal{G}) \leq \left\lfloor \frac{m}{n} \right\rfloor + 1$$

Proof: See Theorem A.5 in Appendix A. □

Corollary 2.37 *If $\mathcal{G}$ is a balanced graph, then*

$$0 \leq a_1(\mathcal{G}) \leq \frac{n}{n-1}\delta_o \leq \frac{n}{n-1}\Delta_o \leq b_1(\mathcal{G}) \leq 2\Delta_o$$

The following result shows that a union of imploding star graphs maximizes the value of a_1 among graphs with the same number of vertices and edges.

Theorem 2.38 *[Cao and Wu (2007)] Consider the set Q of simple graphs with n vertices and $k(n-1)$ edges. Let $\mathcal{G}_k \in Q$ be a union of imploding star graphs, defined as the graph where there exists k vertices such that there is an edge from every other vertex to these k vertices (Fig. 2.7). Then $a_1(\mathcal{G}_k) = k$ is maximal among the elements of the set Q.*

Proof: From Theorem 2.36(11), we know that for $\mathcal{G} \in Q$,

$$a_1(\mathcal{G}) \leq \left\lfloor \frac{k(n-1)}{n} \right\rfloor + 1 = k - 1 + 1 = k$$

The value of a_1 of an imploding star graph (Fig. 2.6) is 1 [Wu (2005b)]. Since $\mathcal{G}_k$ is the union of k imploding star graphs, By Theorem 2.35(6) $a_1(\mathcal{G}_k) \leq k$. □

In Chapter 6 we will show that the union of imploding star graphs also maximizes the ergodicity coefficient. In Section A.3 of Appendix A we will see that a union of $\lfloor \frac{n}{2} \rfloor$ imploding star graphs minimizes the bisection width, isoperimetric number and the minimum ratio cut among all digraphs with n vertices.

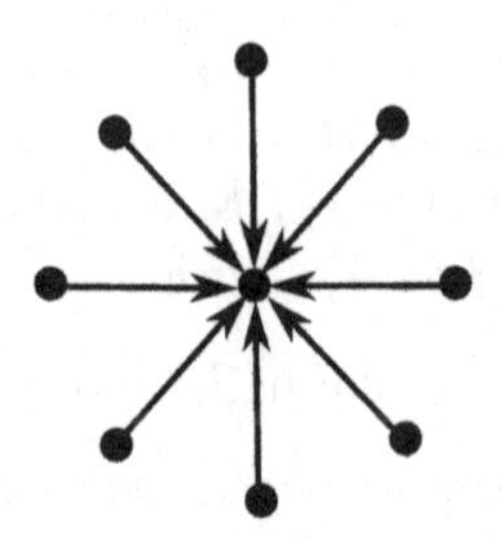

Fig. 2.6 An imploding star graph has a single vertex with an edge from every other vertex pointing to it.

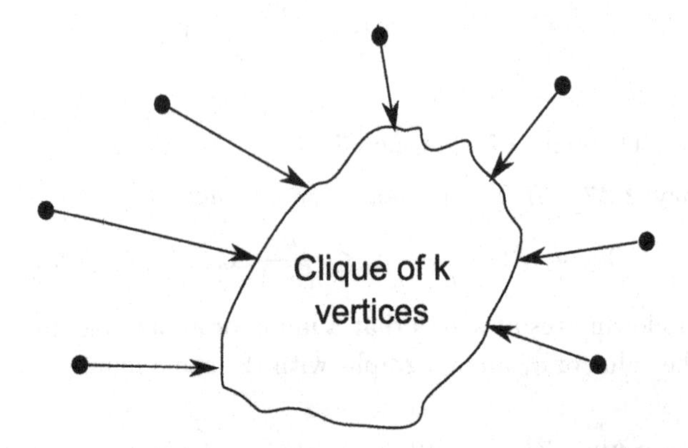

Fig. 2.7 A union of k imploding star graphs. This can be visualized as edges of every vertex ending at vertices in a clique of k vertices. Each arrow in the figure corresponds to k edges ending in each of the k vertices in the clique.

2.3.4 *Basic properties of* a_2

Theorem 2.39

(1) *a_2 can be efficiently computed as*

$$a_2(L) = \min_{x \in \mathbb{R}^{n-1}, \|Qx\|=1} x^T Q^T W L Q x = \lambda_{\min}\left(\frac{1}{2} Q^T \left(WL + L^T W\right) Q\right)$$

where Q is an n by $n-1$ matrix whose columns form an orthonormal basis of $\mathbf{1}^{\perp}$;

(2) *If the reversal of the graph does not contain a spanning connected tree, then $a_2(L) \leq 0$;*

(3) *If the graph is strongly connected, then $0 < a_2 \leq Re(\lambda)$ for all eigenvalues λ of L not belonging to the eigenvector $\mathbf{1}$;*

(4) If $\mathcal{G}$ is strongly connected, then

$$a_2(\mathcal{G}) \geq \frac{1-\cos(\frac{\pi}{n})}{r} e(\mathcal{G})$$

$$a_2(\mathcal{G}) \geq \frac{C_1 e(\mathcal{G})}{2r} - C_2 q$$

where $r = \frac{\max_v w(v)}{\min_v w(v)}$, $q = \max_v w(v) d_o(v)$, $C_1 = 2(\cos(\frac{\pi}{n}) - \cos(\frac{2\pi}{n}))$, $C_2 = 2\cos(\frac{\pi}{n})(1-\cos(\frac{\pi}{n}))$ *and* $e(\mathcal{G})$ *is the edge connectivity (Definition A.12).*

Proof: See Theorem A.6 in Appendix A. □

2.4 Locally connected graphs

Intuitively, a locally connected graph is a graph where the vertices are located on some topological space and each vertex is only connected to a subset of vertices in a neighbourhood.

Definition 2.40 [Wu (2001)] A graph is locally connected if the vertices are arranged on a grid and each vertex is adjacent only to vertices in a local neighborhood of fixed radius. More specifically, a graph is called locally connected with parameters d and k if

(1) The vertices are located on an integer lattice $\mathbb{Z}^d$.
(2) If vertex i is connected to vertex j, then the distance between i and j is less than k.

Theorem 2.41 *Consider a sequence of graphs $\mathcal{G}_i$ with bounded vertex degree where $\mathcal{G}_n$ has n vertices, diameter d_n and Laplacian matrix L_n. If $\limsup_{n\to\infty} \frac{\ln n}{d_n} = 0$, then $\lim_{n\to\infty} \lambda_2(L_n) = 0$.*

Proof: First note that by hypothesis, $2(d_n - 2) - \ln(n-1) > 0$ for large enough n. In [Mohar (1991a)] it was shown that $d_n \leq 2\left\lceil \frac{\Delta+\lambda_2}{4\lambda_2} \ln(n-1) \right\rceil$, where Δ is the maximum vertex degree. This implies that $\lambda_2 \leq \frac{\Delta \ln(n-1)}{2(d_n-2)-\ln(n-1)}$ if $2(d_n-2) - \ln(n-1) > 0$ and the conclusion follows. □

A locally connected graph indicates that the distance between arbitrary vertices are large and this resulted in a small algebraic connectivity.

Corollary 2.42 *Consider a sequence of locally connected graphs $\mathcal{G}_i$ where $\mathcal{G}_n$ has n vertices and Laplacian matrix L_n. If the parameters d and k in Definition 2.40 are bounded, then* $\lim_{n\to\infty} \lambda_2(L_n) = 0$.

Proof: Follows from Theorem 2.41 and the fact that the diameter d_n grows on the order of $\sqrt[d]{n}$. □

On the other hand, the converse of Theorem 2.41 is not true and a small diameter does not guarantee that the algebraic connectivity is large. For instance, consider the "barbell" graph (Fig. 2.8) in [Lin and Zhan (2006)] where two identical complete graphs of n vertices are connected via a single edge. The diameter is 3 and the algebraic connectivity vanishes as $o\left(\frac{1}{n}\right)$.

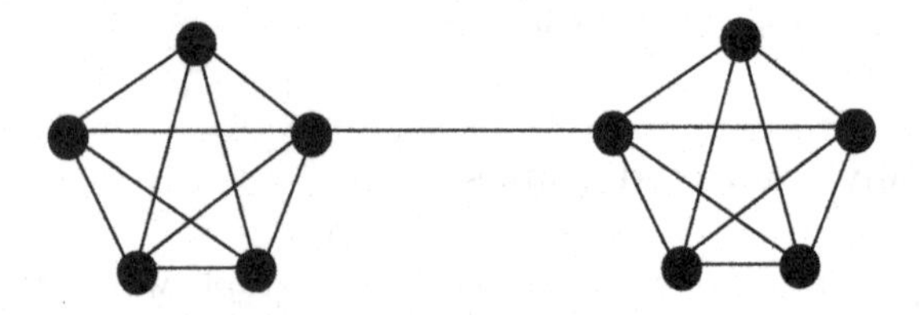

Fig. 2.8 A barbell graph constructed by connecting two disjoint cliques with a single edge.

The following results extend Theorem 2.41 to digraphs.

Theorem 2.43 *[Chung* et al. *(1994)] Let L be the Laplacian matrix of a graph with n vertices and diameter d. If ρ and λ are such that $1 \geq \rho > \|I - \lambda L - \frac{1}{n}\mathbf{1}\mathbf{1}^T\|$, then*

$$d \leq \left\lceil \frac{\ln(n-1)}{\ln \frac{1}{\rho}} \right\rceil$$

Theorem 2.44 *If $\mathcal{G}$ is strongly connected graph with diameter d and w is a positive vector such that $w^T L = 0$ and $\|w\|_\infty = 1$, then*

$$d \leq \left\lfloor \frac{\ln(n-1)}{\ln \frac{\|WL\|}{\sqrt{\|WL\|^2 - a_2^2}}} \right\rfloor + 1$$

where $W = diag(w)$ and $a_2 = a_2(L)$.

Proof: The proof is similar to the proof of Theorem 6.3 in [Chung *et al.* (1994)]. □

Corollary 2.45 *Consider a sequence of digraphs $\mathcal{G}_i$ where $\mathcal{G}_n$ has n vertices, diameter d_n and Laplacian matrix L_n. If $\lim_{n\to\infty} \frac{\ln n}{d_n} = 0$, then $\lim_{n\to\infty} \lambda_2(L_n) = 0$.*

Corollary 2.46 *Let G_i be a sequence of strongly and locally connected (di)graphs, then $\lim_{n\to\infty} a_2(L_n) = 0$.*

In Section 3.5 we look at a class of random graphs that can also be considered to be locally connected.

2.4.1 *Basic properties of a_3 and a_4*

Theorem 2.47

(1) If the graph is strongly connected, then $a_3 \geq a_2 > 0$;
(2) $a_4 > 0$ if and only if the reversal of the graph contains a spanning directed tree.

Proof: See Theorem A.9 in Appendix A. □

As we will see in Chapters 4 and 5, the quantities a_i are useful in deriving synchronization criteria in networks of coupled dynamical systems. The algebraic connectivity was originally proposed as a measure to quantify the connectedness of the graph. There exist connections between the algebraic connectivity and various combinatorial properties of the graph such as diameter, isoperimetric number and minimum ratio cut. We have summarized some of these relationships in Appendix A. This shows how properties of the coupling graph can affect the ability of the network to synchronize. In general, we see that the more "connected" the underlying graph is, the easier it is to synchronize the network.

2.5 Examples

Table 2.5 shows some directed and undirected graphs along with their values of a_4 and b_1.

By the super-additivity property of a_1 (Theorem 2.35), adding extra undirected edges to a graph does not decrease a_1. Therefore a_1 of the undirected cycle graph[6] is not smaller than a_1 of the undirected path graph. This is not true if we add *directed* edges. For example, a_4 of the directed path graph (Fig. 2.9(e)) is 1 independent of the number of vertices n, but

[6] Which is equal to a_4 by Lemma 2.34.

Graph of order n	a_4	b_1
undirected complete (Fig. 2.9(a))	n	n
undirected star (Fig. 2.9(b))	1	n
undirected path (Fig. 2.9(c))	$2+2\cos\left(\frac{(n-1)\pi}{n}\right)$	$2+2\cos\left(\frac{\pi}{n}\right)$
undirected cycle (Fig. 2.9(d))	$4\sin^2\left(\frac{\pi}{n}\right)$	$4\sin^2\left(\frac{\lfloor\frac{n}{2}\rfloor\pi}{n}\right)$
directed path (Fig. 2.9(e))	1	
directed cycle (Fig. 2.9(f))	$1-\cos\left(\frac{2\pi}{n}\right)$	

Table 2.5: Table of graphs and their corresponding values of a_4 and b_1.

by adding a single directed edge, the resulting directed cycle graph (Fig. 2.9(f)) has a_4 converging to 0 as $n \to \infty$. This is an important difference between directed and undirected graphs. This phenomena is reminiscent of the Braess paradox [Braess (1968)] in traffic routing, where the addition of roads can lead to more traffic congestion. From a synchronization point of view, the main difference between Fig. 2.9(e) and Fig. 2.9(f) is that the strongly connected components of the directed path graph are single vertices and thus are easy to synchronize, i.e. the first two vertices are synchronized, then the next vertex is synchronized to the first two, and so forth, i.e. synchronization is reached in stages, each time considering a subgraph of two vertices. On the other hand, in the directed cycle graph the strongly connected component is the entire graph with a large diameter resulting in each vertex influencing and being influenced by vertices which are far apart and thus making the network harder to synchronize.

What this means is that under appropriate conditions adding additional symmetric coupling to a network coupled with symmetric coupling will not decrease the synchronizability of the network. However in a network with unidirectional coupling, adding additional coupling can lead to a loss of synchronization, a conclusion which first appears counterintuitive.

2.6 Hypergraphs

Hypergraphs are generalization of (undirected) graphs. Edges in graphs are 2-element subsets of vertices. Edges in a hypergraph, called hyperedges, are nonempty subsets of vertices of at least 2 elements. A vertex v is incident to a hyperedge e if $v \in e$. The incidence matrix of a hypergraph with n

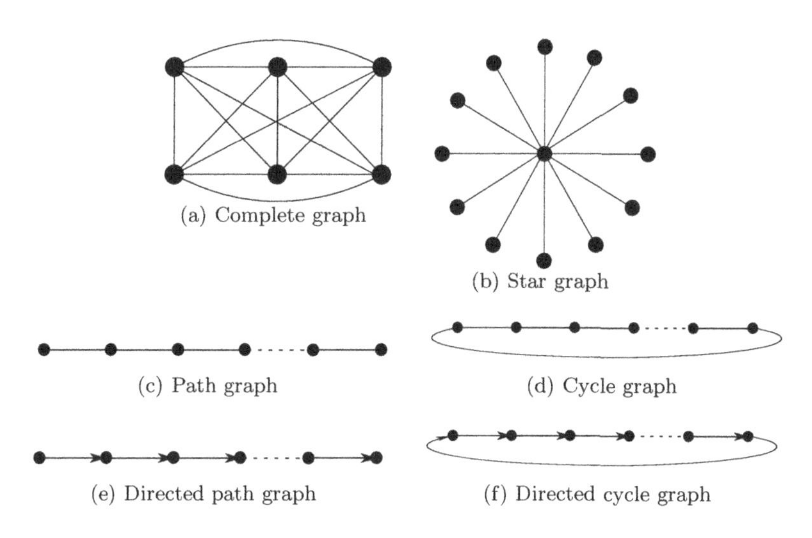

(a) Complete graph

(b) Star graph

(c) Path graph

(d) Cycle graph

(e) Directed path graph

(f) Directed cycle graph

Fig. 2.9 Undirected and directed graphs

vertices and m edges is defined as the n by m $0-1$ matrix M where $M_{ij} = 1$ if and only if vertex i is incident to the j-th edge. The Laplacian matrix of a hypergraph with incidence matrix M can be defined as [Bolla (1993); Wu (1998b)]:

$$L = 2(D_v - MD_e^{-1}M^T)$$

where D_v and D_e are diagonal matrices with the degrees of the vertices and edges respectively. This definition corresponds to the definition of Laplacian matrix for graphs if we map a hypergraph to a graph by replacing each hyperedge of k vertices with a clique of k vertices where each edge has weight $\frac{2}{k}$.

2.7 Further reading

The reader is referred to [Swamy and Thulasiraman (1981); Bollobás (2001); Godsil and Royle (2001)] for general references on graph theory. Theorem 2.23 and Corollary 2.25 have been shown independently using different techniques in at least the following publications [Fax (2002); Ren and Stepanyan (2003); Agaev and Chebotarev (2005); Wu (2005c); Lin *et al.* (2005); Lafferriere *et al.* (2005)]. The various generalizations of the concept of algebraic connectivity to digraphs can be found in [Wu (2005b,d,g)].

Chapter 3

Graph Models

There exists many types of complicated networks, some man-made, others naturally occurring. The main goal in this book is to study coupled dynamical systems where the coupling is expressed via such networks. In this chapter we briefly describe examples of such networks and ways to generate them computationally and characterize them mathematically.

3.1 Examples of complex networks

3.1.1 *Neural networks*

The most important component of our central nervous system is an incredibly complicated interconnected network of neurons that is responsible for enabling every function of our body. The main components of a neuron are the body (also called the soma), the axon, and the dendrites. The end of the axon consists of axon terminals that connect via synapses to dendrites of other neurons. When a neuron fires, a pulsed electrical signal is generated and propagates along the axon to synapses that transfer the information in the signal to other neurons by releasing chemicals called neurotransmitters across the synaptic gap. This is shown schematically in Fig. 3.1. It is estimated that there are close to 100 billion neurons in the human body. On average, each neuron has thousands of synapses. The resulting extremely intricate neural network is responsible for regulating autonomic functions such as the beating of the heart and maintaining balance, but is also responsible for our ability to solve Sudoku puzzles, play Go and write Haiku. There is evidence that other types of cells besides neurons also participate in information processing in the brain. This is obviously a very simplified view of how a biological neural network functions. The reader is referred to [Kolb and Whishaw (1990)] for more details on biological neurons and the central nervous system.

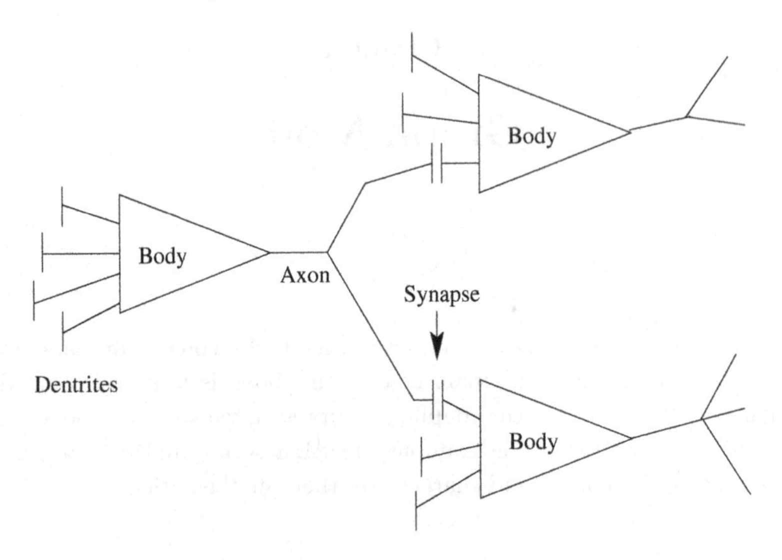

Fig. 3.1 Schematic diagram of a neural network.

3.1.2 *Transportation networks*

An excellent example of a man-made network is the transportation network, for instance a network of roads connecting cities, or the network of (passenger or freight) airplane flights between cities. Analyzing and understanding such networks is of great utility to modern society and several problems regarding such networks are notoriously hard. For instance, in the Traveling Salesman Problem (TSP) which is NP-complete, the goal is to find the shortest path which starts at one city, visits all cities once and return to the starting city. In recent years, significant strides have been made towards solving moderately sized TSP problems. In 2004, the TSP problem for all the cities in Sweden (about $25,000$) was solved. There are currently efforts underway in solving the TSP problem for the almost 2 million cities in the world. As of July 2006, the best path found so far is less than 0.07% longer than the shortest path! (`http://www.tsp.gatech.edu/world/index.html`). In [Guimerà *et al.* (2005)] it was found that the world wide air transportation network contains features of both small-world networks (Section 3.3) and scale-free networks

(Section 3.4). The properties of river networks is studied in [Banavar *et al.* (1999)].

3.1.3 *World Wide Web*

A more recent example of man-made networks is the networks of webpages comprising the World Wide Web. The vertices are webpages that exist in the internet and there is a directed edge from webpage A to webpage B if there is a hypertext link in A that points to B. Obviously the topology of this graph changes with time and during the late 1990's the number of vertices and edges were growing at a rapid rate. Some properties of this graph has been analyzed in [Kumar *et al.* (2000)] where it was shown that this web graph in 1999 has a "bowtie" structure where almost the entire graph consists of a strongly connected component and two sets of vertices which point into and out of this component respectively.

3.1.4 *Erdös number and scientific collaboration network*

The vertices in the scientific collaboration network are individuals and two persons are connected if they have published a scientific paper together, i.e. the edges in this undirected graph are scientific papers. The Erdös number of a person is the length of the shortest path, (i.e. the distance) between the mathematician Paul Erdös and that person. You can find more information about this graph at `http://www.oakland.edu/enp/`. Readers with access to MathSciNet of the American Mathematical Society can find out their Erdös number (or at least an upper bound since the database does not contain all scientific papers) at `http://www.ams.org/mathscinet/`. The shortest path which links a person to Erdös may not be unique. As of this writing my Erdös number is 4 since I have written a paper with Charles Tresser [Balmforth *et al.* (1997)] who has written a paper with Jaume Llibre [Alsedà *et al.* (1989)] who has written a paper with Branko Grünbaum [Artés *et al.* (1997)] who has written a paper with Paul Erdös [Erdös and Grünbaum (1973)].

3.1.5 *Film actor network*

This undirected graph is similar to the Erdös number graph. The vertices in this graph are actors and actresses around the world and the edges are motion pictures. Two actors or actresses are connected if they have appeared together in a film. A popular game on U.S. college campuses (at some point in time at least) is to find a path connecting the actor Kevin Bacon to an arbitrary actor or actress. The length of the shortest path (the Kevin Bacon number) would surprise you; it is usually smaller than you think. Some properties of this graph (which, as in the Erdös graph, naturally changes over time) can be found at `http://www.cs.virginia.edu/oracle/` and `http://www.cs.virginia.edu/oracle/star_links.html` which also allow you to find the shortest path between any two actors/actresses. For instance, the mathematician Brian Greene, who has an Erdös number 3 [Greene *et al.* (1990); Chung *et al.* (1996); Erdös and Graham (1972)] has a Kevin Bacon number 2 since he appeared in the film Frequency (2000) with John Di Benedetto who appeared in Sleepers (1996) with Kevin Bacon[1].

Social networks such as the Erdös numbers graph and the Kevin Bacon graph are more naturally expressed as hypergraphs, i.e. graphs where the edges are subsets of vertices. For example, in the Erdös number graph, a hyperedge is the set of co-authors in a paper.

3.1.6 *Citation network*

In this network, the vertices are scientific papers with a directed edge from paper A to paper B if and only if paper A cites paper B. Price [de Solla Price (1965, 1976)] studied this type of networks through a model of cumulative advantage in which paper which are cited often are more likely to cited, and shows that both the in-degree distribution and the out-degree distribution follow a distribution related to Legendre's beta function[2] which is well approximated by a power law distribution.

In order to model these complicated networks, many random graph models have been proposed. In the remainder of this chapter we examine some of these models along with properties of these graphs useful in our study of synchronization properties.

---

[1] The sum of the Erdös number and the Kevin Bacon number of an individual is called the Erdös-Bacon number.

[2] This is also known as the Eulerian integral of the first kind and is defined as $\beta(a,b) = \frac{(a-1)!(b-1)!}{(a+b-1)!}$.

3.2 Classical random graph models

These are the original models of *random* graphs first defined and studied extensively by Erdös and Rényi [Erdös and Renyi (1959)] and by Gilbert [Gilbert (1959)] around 1959. For a given n, these models define a probability measure on the space of all graphs of n vertices. In particular, the random graph model $G_1(n, M)$ with $N = \binom{n}{2}$ and $0 \leq M \leq N$ is defined as the space of all graphs with n vertices and M edges, each one of them equiprobable with probability $\binom{N}{M}^{-1}$. The random graph model $G_2(n, p)$ with $0 \leq p \leq 1$ is defined as graphs where each edge is chosen with probability p. Thus a graph with m edges will have a probability $p^m(1-p)^{N-m}$.

In general the parameters M and p are functions of n. When $M \approx pN$, the two models $G_1(n, M)$ and $G_2(n, p)$ are essentially equivalent for large n [Bollobás (2001)].

The random graph models $G_1(n, M)$ and $G_2(n, p)$ are easily extended to digraphs. In particular, we define the random digraph model $G_{d1}(n, M)$ as follows. Each digraph of n vertices and M directed edges is given equal probability whereas other graphs of n vertices has probability 0.

A digraph of n vertices has $n(n-1)$ possible directed edges and for the random digraph model $G_{d2}(n, p)$ each edge is chosen with probability p.

3.2.1 *Algebraic connectivity of random graphs*

In this section we study some properties of the algebraic connectivity of these models. We first start with the regular random graph model $G_{1r}(n, M)$, where every regular graph with n vertices and M edges has equal probability and all other graphs has probability 0.

Theorem 3.1 *Consider the 2d-regular random graphs $G_{1r}(n, 2d)$. For n and d large, with high probability the eigenvalues of the Laplacian matrix satisfy*

$$\lambda_2 > 2d - 2\sqrt{2d-1} - 2\log d$$

$$\lambda_n < 2d + 2\sqrt{2d-1} + 2\log d$$

$$\frac{\lambda_n}{\lambda_2} \leq 1 + \frac{2\sqrt{2d-1} + 2\log d}{d - \sqrt{2d-1} - \log d}$$

Proof: Follows from Theorem C in [Friedman *et al.* (1989)]. □

Another class of regular graphs with similar eigenvalue properties are the Ramanujan graphs [Lubotzky *et al.* (1988); Murthy (2003)]. These k-regular graphs possess the following property: the second smallest eigenvalue of the Laplacian matrix satisfies

$$\lambda_2 \geq k - 2\sqrt{k-1}$$

Explicit construction of Ramanujan graphs for a fixed k are only known when $k-1$ is a prime power. In [Miller *et al.* (2006)] results are presented that suggest that for many classes of regular graphs, over 52% of them should be Ramanujan graphs.

Theorem 3.2 *For any $\epsilon > 0$, the algebraic connectivity of graphs in $G_{1r}\left(n, \frac{kn}{2}\right)$ is larger than $\left(1 - \frac{\sqrt{3}}{2} - \epsilon\right)k$ almost everywhere as $n \to \infty$ for a large enough k. For even k, the algebraic connectivity is close to k (relative to the magnitude of k) almost everywhere as $n \to \infty$.*

Proof: According to [Bollobás (1988)] the isoperimetric number of almost every k-regular random graph satisfies $i \geq \frac{k}{2} - \sqrt{k\ln 2}$ for large enough k. According to [Mohar (1989)] the algebraic connectivity a_1 of a graph satisfies $i \leq \sqrt{a_1(2k-a_1)}$. Combining these two inequalities results in

$$a_1 \geq k - \sqrt{\frac{3}{4}k^2 - k\left(\ln 2 - \sqrt{k\ln 2}\right)}$$

The second part follows from Theorem 3.1. □

For small k, these arguments can be used to show that for $k \geq 3$, $a_1 \geq 0.0018k$ almost everywhere as $n \to \infty$. In [Juhász (1991)] the algebraic connectivity of the random graph model $G_2(n,p)$ is calculated to be:

Theorem 3.3 *For a graph $\mathcal{G}$ in $G_2(n,p)$, $\lambda_2(\mathcal{G})$ satisfies*

$$\lambda_2(\mathcal{G}) = pn + o\left(n^{\frac{1}{2}+\epsilon}\right) \text{ in probability.}$$

for any $\epsilon > 0$.

For large n, $G_2(n,p)$ has average degree around pn, i.e. it has a high probability that there is a vertex with degree at most pn. Since $\lambda_2 \leq pn$ by Theorem 2.31(3), this implies that random graphs have the largest possible

λ_2 (on a relative basis as a function of n) among graphs with the same average degree.

Theorem 3.3 can be extended to random digraphs $G_{d2}(n,p)$:

Theorem 3.4 *[Wu (2005b)] For any $\epsilon > 0$, the algebraic connectivity of a graph $\mathcal{G}_d$ in $G_{d2}(n,p)$ satisfies*

$$a(\mathcal{G}_d) = pn + o\left(n^{\frac{1}{2}+\epsilon}\right) \text{ in probability.}$$

Proof: The proof is analogous to the proof in [Juhász (1991)]. Consider the symmetric matrix $B = \frac{1}{2}(A + A^T)$. Since $P(B_{ij} = 0) = q^2$, $P(B_{ij} = \frac{1}{2}) = 2pq$, $P(B_{ij} = 1) = p^2$, by [Füredi and Komlós (1981)],

$$\max_{i \leq n-1} |\lambda_i(B)| = o(n^{\frac{1}{2}+\epsilon}) \text{ in probability.}$$

Let $C = \frac{1}{2}(L + L^T) - (D_B - B)$ where D_B is the diagonal matrix with the row sums of B on the diagonal. Note that $\mathbf{1}$ is an eigenvector of $D_B - B$ and thus $\min_{x \in P} x^T(D_B - B)x = \lambda_2(D_B - B)$. Consider the diagonal matrix $F = (D_B - p(n-1)I)$. As in [Juhász (1991)] the interlacing properties of eigenvalues of symmetric matrices implies that $|\lambda_2(D_B - B) - \lambda_2(p(n-1)I - B)| \leq \rho(F) \leq \|F\|_\infty$. An application of a generalization of Chernoff's inequality [Alon and Spencer (2000)] (also known as Hoeffding's inequality) shows that $P(\|F\|_\infty \geq Kn^{\frac{1}{2}+\epsilon}) \leq \sum_i P(|F_{ii}| \geq Kn^{\frac{1}{2}+\epsilon}) \leq ne^{-\beta n^{2\epsilon}}$ and thus $\|F\|_\infty = o(n^{\frac{1}{2}+\epsilon})$ in probability. Therefore $|\lambda_2(D_B - B) - p(n-1) + \lambda_{n-1}(B)| = o(n^{\frac{1}{2}+\epsilon})$, i.e. $|\lambda_2(D_B - B)| = pn + o(n^{\frac{1}{2}+\epsilon})$. Next note that $C = D - D_B$ is a diagonal matrix and

$$a(\mathcal{G}_d) = \min_{x \in P} x^T Lx \leq \lambda_2(D_B - B) + \max_{x \in P} x^T Cx \leq \lambda_2(D_B - B) + \max_i C_{ii}$$

Similarly,

$$a(\mathcal{G}_d) \geq \lambda_2(D_B - B) + \min_{x \in P} x^T Cx \geq \lambda_2(D_B - B) + \min_i C_{ii}$$

i.e. $|a(\mathcal{G}_d) - \lambda_2(D_B - B)| \leq \|C\|_\infty$. Similar applications of Hoeffding's inequality show that $\|C\|_\infty = o(n^{\frac{1}{2}+\epsilon})$ in probability which implies that $\|F\|_\infty + \|C\|_\infty = o(n^{\frac{1}{2}+\epsilon})$ in probability and thus the theorem is proved. □

3.3 Small-world networks

A small but well-known study carried out by Milgram in the 1960's where letters are sent from one person to another by passing them though acquaintances show that the number of acquaintances connecting one person to another is generally small [Milgram (1967)][3]. This is referred to as the *small world* phenomenon and spurred the phrase "six degrees of separation" which was coined by John Guare in his play of the same name. There was an experiment conducted to test the same concept using messages sent through the internet, but as of this writing the website seems to be closed. Recently several models have been proposed to algorithmically generate networks capturing the small world property.

3.3.1 Watts-Strogatz model

The small world effect can be characterized by a small diameter of the graph. One way the diameter can be made small is by introducing random edges. In the model proposed in [Watts and Strogatz (1998)], the random graph is generated iteratively starting from a graph with only nearest neighbor edges. At each iteration, edges in the graph with only nearest neighbors connections are *replaced* with random edges with a probability p.

3.3.2 Newman-Watts model

The model in [Newman and Watts (1999)] is a slight modification of the Watts-Strogatz model, where random edges are *added* with probability p to a graph with nearest neighbors connections.

Examples of these two models of small world networks are shown in Figure 3.2.

3.3.3 Algebraic connectivity of small-world networks

In the small world network models above, a small world network is essentially a union of two types of graphs, a locally connected graph and a random graph. The superadditivity of λ_2 (Theorem 2.35(6)) allows us to show that λ_2 of a small world network is dominated by λ_2 of the random component as $n \to \infty$. More precisely, the Laplacian matrix L of a small world network after k random edges are added can be decomposed as $L = L_l + L_r$, where L_r is Laplacian matrix of a member of $G_1(n, k/N)$, where $N = \binom{n}{2}$. L_l is the Laplacian matrix of the nearest neighbor cycle

[3]The validity of the experiment has been questioned, but this experiment is the first to show the possibility of a social network possessing small world characteristics.

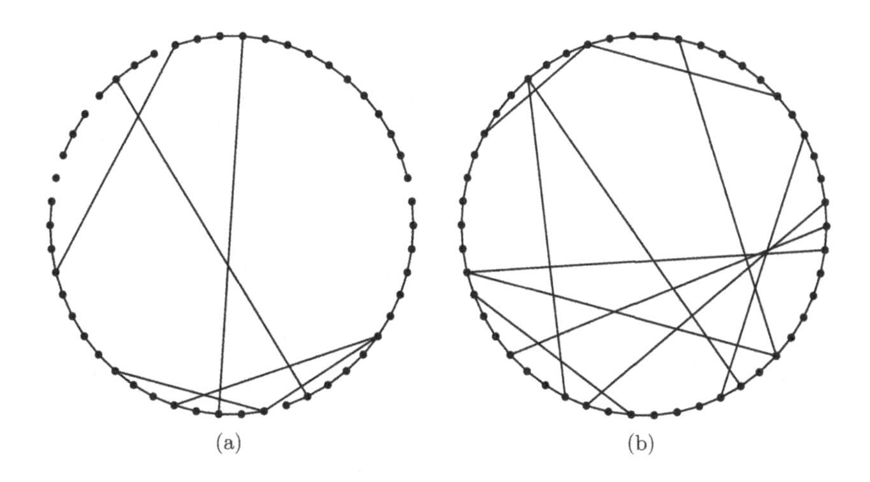

Fig. 3.2 Small world networks. (a) Watt-Strogatz model. (b) Newman-Watts model.

graph (Fig. 2.9(d)) for the Newman-Watts model and the Laplacian matrix of a subgraph of the nearest neighbor graph for the Watts-Strogatz model. In either case $\lambda_2(L_l) \to 0$ as $n \to \infty$ by Corollary 2.42 since this graph is locally connected. On the other hand, $\lambda_2(L_r)$ is bounded away from 0 as $n \to \infty$ (Theorem 3.3) and approach $\frac{2k}{n-1}$ for large k and n. In other words, $\lambda_2(L) \approx \lambda_2(L_r) \approx \frac{2k}{n-1}$ as $k, n \to \infty$.

For a sequence of undirected graphs with increasing number of vertices n and constant average degree m, let us define a *connectivity measure* $s(m)$ as $s(m) = \lim_{n\to\infty} \frac{\lambda_2}{m}$. By Corollary 2.37 $\frac{\lambda_2}{m} \le \frac{n}{n-1}$, and thus $0 \le s(m) \le 1$.

After k edges are added to a nearest neighbour cycle graph, the average vertex degree is $m = 2 + \frac{2k}{n}$ and $\frac{\lambda_2}{m} \approx \frac{m-2}{m}\frac{n}{n-1}$ for large k and n. Let $0 \le p \le 1$ be the fraction of edges that are added in the Newman-Watts model at each iteration. Then $m = 2 + 2p$ and thus $s(m) \approx \frac{p}{p+1}$. Similarly, for the Watts-Strogatz model, $m = 2$ and $s(m) \approx p$.

3.4 Scale-free networks

The degree distribution of a graph is the histogram of the sequence of vertex degrees. Scale-free networks are characterized by the degree distribution following a power law. In other words, the number of vertices with degree k is proportional to k^d for some exponent d. One way this type of graphs is generated is through the preferential attachment process proposed in [Barabási *et al.* (2000)] where edges with high degrees have a higher probability to connect with new vertices as they are added to the graph. In this process, a graph is generated iteratively by adding at each iteration a vertex and edges from this vertex to the vertices with the highest degrees. A similar process has been proposed by de Solla Price (Sec. 3.1.6) earlier in his study of citation networks and in fact the power law is an approximation of the distribution obtained in [de Solla Price (1965, 1976)].

For scale free networks, the latest vertex added has degree $k = \frac{m}{2}$ and thus the connectivity measure satisfies $s(m) \leq \frac{1}{2}$. Computer simulations[4] in [Wu (2003a)] show that for scale free networks for fixed m, λ_2 increases monotonically as n increases. Fig. 3.3 shows that $s(m)$ increases monotonically as a function of m and approaches the upper bound $\frac{1}{2}$. We will argue that $s(m)$ does indeed approach $\frac{1}{2}$ as $n \to \infty$ using a prescribed degree sequence model in Section 3.6.

3.5 Random geometric graphs

In this model [Gilbert (1961); Dall and Christensen (2002); Penrose (2003)], vertices are randomly chosen in a d-dimensional unit cube and two vertices are connected by an edge if and only if their distance is smaller than a fixed number r. An example of such a graph instance with 100 vertices and $d = 2$, $r = 0.1$ is shown in Fig. 3.4.

This model is useful to model for instance a collection of wireless devices that can only communicate with another device if they are close enough to each other (see the examples in Chapter 7). If the vertices are chosen in the unit ball the following was shown:

Theorem 3.5 *[Ellis* et al. *(2006)] The diameter D of a geometric graph*

[4]The construction in [Wu (2003a)] differs slightly from [Barabási *et al.* (2000)] in the construction of the scale-free networks. In [Barabási *et al.* (2000)] the initial configuration is k_0 vertices with no edges. After many iterations, this could still result in some of these initial k_0 vertices having degree less than k, even though every newly added vertex has degree at least k. Since even one vertex with degree 1 results in λ_2 to drop below 1, we would like to have every vertex of the graph to have degree at least k. Therefore we will start with the initial k_0 vertices fully connected to each other for the case $k_0 = k$. For the case $k_0 > k$, we require that the initial k_0 vertices are connected with each other with vertex degrees at least k. This guarantees that the vertex degree of every vertex is at least k.

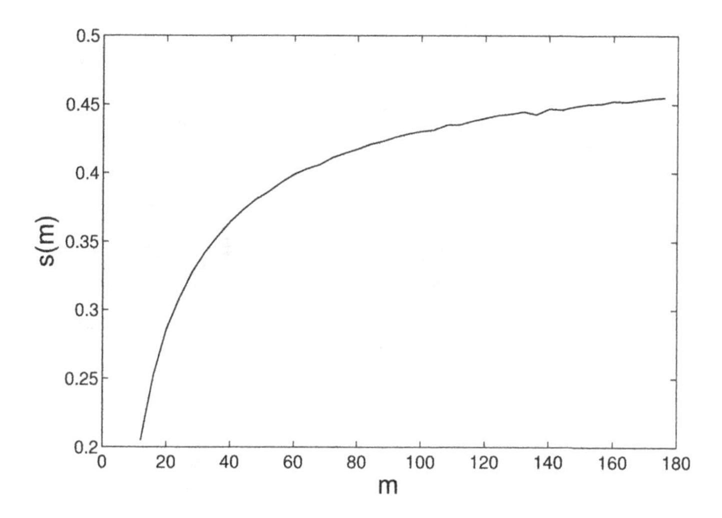

Fig. 3.3 Connectivity measure $s(m)$ versus m of scale free networks obtained from computer simulations.

in the d-dimensional unit ball satisfies

$$\frac{1-o(1)}{r} \leq D \leq \frac{1+O((\ln(\ln(n))/\ln(n))^{\frac{1}{d}})}{r}$$

almost always.

This in combination with Theorem 2.41 shows that the algebraic connectivity vanishes as $n \to \infty$ if for a sequence of graphs, r as a function of n decreases faster than $\frac{1}{\ln(n)}$.

3.6 Graphs with a prescribed degree sequence

The degree sequence of a graph is a list of the degrees of the vertices. A sequence of nonnegative integers is called *graphical* if it is the degree sequence of some simple graph. An explicit sufficient and necessary condition for a sequence to be graphical is first given in [Erdös and Gallai (1960)]:

Theorem 3.6 *A sequence of nonnegative integers* $(d_1, \cdots, d_n)$ *is graph-*

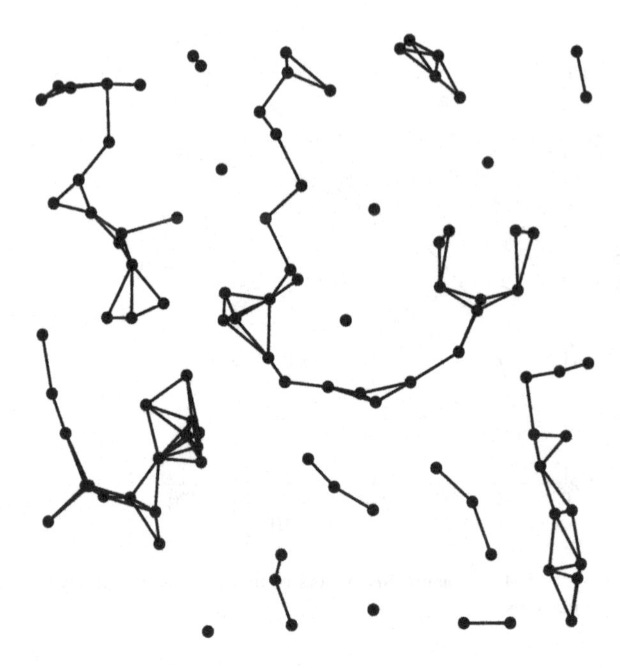

Fig. 3.4 An instance of a random geometric graph with 100 vertices in the unit square and vertices are connected if and only if their distance is less than 0.1.

ical if and only if

$$\sum_{i=1}^{k} d_i \leq k(k-1) + \sum_{j=k+1}^{n} \min(k, d_j)$$

for each $1 \leq k \leq n$.

Other equivalent criteria for a sequence to be graphical can be found in [Sierksma and Hoogeveen (1991)]. A constructive criterion for a sequence to be graphical is the following [Havel (1955); Hakimi (1962)]:

Theorem 3.7 *A sequence of nonnegative integers* $(d_1, \cdots, d_n)$ *is graphical if and only if the sequence obtained by removing the largest entry* d_i *from the list and subtracting 1 from the largest* d_i *of the remaining entries is graphical.*

Theorem 3.7 provides the following algorithm to construct a graph given a graphical sequence $(d_1, \cdots, d_n)$. Starting with the list $(d_1, \cdots, d_n)$ the vertex with the largest d_i is connected to d_i other vertices with the largest degrees. This list is updated by removing the largest entry d_i and subtracting 1 from the largest d_i elements of the list and the above steps are repeated.

Consider the following random undirected graph model in [Wu (2003a)]. Given a graphical sequence $d = (d_1, \cdots, d_n)$, the random graph model $G_{pd}(d)$ is constructed by choosing an edge between vertex i and vertex j with the following probability:

$$p_{ij} = \frac{d_{\min}}{n} + \frac{(d_i - d_{\min})(d_j - d_{\min})}{\sum_k (d_k - d_{\min})} \tag{3.1}$$

where $d_{\min} = \min_i d_i$.[5]

For computational convenience, this graph model will have graphs with self-loops. Since $\sum_i p_{ij} = d_j$, the graph will have the desired degree sequence in expectation[6]. Since $p_{ij} \geq \frac{d_{min}}{n}$, this model has the benefit that the graphs in the random graph model $G_2(n,p)$ with $p = \frac{d_{\min}}{n}$ can be considered as "subgraphs" of this model. This is made more precise in the following Lemma.

Lemma 3.8 *[Wu (2005e)] Let f be a monotonically nondecreasing function, i.e. if $x \geq y$, then $f(x) \geq f(y)$. Let $E_1 = E(f(\lambda_i(\mathcal{G})))$ for graphs in the model $G_2(n,p)$ where $p = \frac{d_{\min}}{n}$ and let $E_2 = E(f(\lambda_i(\mathcal{G})))$ for graphs in the model $G_{pd}(d)$. Then $E_1 \leq E_2$. In particular, the expected value of algebraic connectivity a_1 in the graph model $G_{pd}(d)$ is larger than or equal to that of $G_2(n,p)$.*

Proof: First we partition the probability space of $G_{pd}(d)$ such that for each graph $\mathcal{H}$ in $G_2(n,p)$ with probability q, there exists exactly one event A in $G_{pd}(d)$ with probability q, such that $\mathcal{H}$ is a subgraph of all the graphs in event A. In particular, consider the edges in $G_{pd}(d)$ to be of two types. An edge (i,j) of type 1 occurs with probability $\frac{d_1}{n}$ and an edge (i,j) of type 2 occurs with probability $P_{ij} - \frac{d_1}{n}$. To each graph $\mathcal{H}$ in $G_2(n,p)$ corresponds a set of graphs $A(\mathcal{H})$ constructed as follows. If (i,j) is an edge in $\mathcal{H}$, then (i,j) is an edge of type 1 in graphs in $A(\mathcal{H})$. If (i,j) is not an edge in $\mathcal{H}$, then either (i,j) is not an edge or (i,j) is an edge of type 2 in graphs in $A(\mathcal{H})$. It is clear that these sets $A(\mathcal{H})$ over all graphs $\mathcal{H}$ in $G_2(n,p)$ exactly partition the probability space of $G_{pd}(d)$. Furthermore, the probability of

[5] If $d_1 = d_2 = \cdots = d_n$, then p_{ij} is defined as $\frac{d_1}{n}$ for all i, j and the resulting graph model is $G_2(n,p)$ with $p = \frac{d_1}{n}$.

[6] This model is a modification of the graph model in [Chung and Lu (2002)] where p_{ij} is proportional to $d_i d_j$.

$A(\mathcal{H})$ in $G_2(n,p)$ is equal to the probability of $\mathcal{H}$ in $G_{pd}(d)$ and $\mathcal{H}$ is a subgraph of every graph in $A(\mathcal{H})$. □

Thus for this graph model we expect $\lambda_2(G) \leq d_1 \frac{n}{n-1}$ by Theorem 2.31. Furthermore, as $p_{ij} \geq \frac{d_1}{n}$, this random graph is equivalent to adding additional edges to the random graph with degree sequence $(d_1, \ldots, d_1)$. Theorem 3.3 and Lemma 3.8 show that $\lambda_2(G) \approx d_{\min}$ for large $d_{\min}$. The average degree is $m = \frac{\sum_{i=1}^n d_i}{n}$, i.e. the connectivity measure $s(m) = \lim_{n\to\infty} \frac{nd_{\min}}{\sum_{i=1}^n d_i}$. For a degree sequence whose distribution follows a power law this is another model for a scale free network. Consider the following power law case where the vertex degree varies from d_1 to d_n with d_1 independent of n and the number of vertices with degree i proportional to $i^{-\alpha}$. It can be shown that in this case $s(m) \approx \lim_{n\to\infty} d_1 \frac{\sum_{i=d_1}^{d_n} i^{-\alpha}}{\sum_{i=d_1}^{d_n} i^{1-\alpha}}$. Approximating the sums by the definite integrals of $x^{-\alpha}$ and $x^{1-\alpha}$ we obtain $s(m) \approx \lim_{n\to\infty} d_1 \frac{(2-\alpha)(d_n^{1-\alpha} - d_1^{1-\alpha})}{(1-\alpha)(d_n^{2-\alpha} - d_1^{2-\alpha})}$. If we assume that $d_n \to \infty$ as $n \to \infty$ (which is the case for the power law model in [Aiello *et al.* (2001)]), this reduces to $s(m) \approx \frac{2-\alpha}{1-\alpha} = 1 - \frac{1}{\alpha-1}$ for $\alpha \geq 2$. It can be shown that $s(m) = 0$ for $\alpha \leq 2$. For $\alpha = 3$, which is the same power law as the scale-free case in Section 3.4, $s(m) \approx 0.5$ when $d_1 \leq m$ is large. This validates the findings illustrated in Fig. 3.3. Note also that as $\alpha \to \infty$, $s(m) \to 1$.

In summary, random graphs and locally connected graphs have the largest and smallest $s(m)$ respectively while scale free and small world networks have connectivity measures somewhere in between. This is summarized in Table 3.1.

Graph class	connectivity measure $s(m)$
Locally connected graphs	0
Newman-Watts	$\approx \frac{p}{1+p}$
Watts-Strogatz	$\approx p$
Scale-free networks	≈ 0.5 as $m \to \infty$
Model in [Wang and Chen (2002)]	$s(m) \approx 1$, $\lambda_2 \to \infty$ as $n \to \infty$
Random graphs	$s(m) \approx 1$, as $m \to \infty$.

Table 3.1: Several classes of graphs and their connectivity measure $s(m)$.

3.7 Algebraic connectivity and degree sequence

Recently, the value of $a_1 = \lambda_2(L)$ and $r = \frac{a_1}{b_1}$ was studied for undirected graphs with a prescribed degree sequence, such as a power law degree sequence. As noted in [Atay *et al.* (2006)] the degree sequence by itself is not sufficient to determine these values. For instance, Theorem 3.3 and Corollary 2.42 show that local coupling and random coupling are at opposite extremes in terms of the network's ability to synchronize.

For a give degree sequence $(d_1, \cdots, d_n)$, we construct in this section two graphs for which a_1 and r differ significantly between the graphs.

3.7.1 *Regular graphs*

3.7.1.1 *Construction 1: graph with low λ_2 and r*

For a $2k$-regular graph, the degree sequence is $(2k, 2k, \ldots, 2k)$. We consider two cases. In the first case, the vertex degree grows as $\Omega\left(n^{\frac{1}{3}-\epsilon}\right)$ for $\epsilon > 0$. If we arrange the vertices in a circle, and connect each vertex to its $2k$ nearest neighbors, then the resulting $2k$-regular graph is denoted C_{2k}. The Laplacian matrix L is a circulant matrix:

$$\begin{pmatrix} 2k & -1 & \cdots & -1 & 0 & \cdots & 0 & -1 \cdots & -1 \\ -1 & 2k & -1 & \cdots & -1 & & & & \\ & \ddots & \ddots & & \ddots & & & & \\ -1 & \cdots & -1 & 0 & \cdots & 0 & -1 & \cdots & -1 \; 2k \end{pmatrix}$$

The eigenvalues of L are given by $(\mu_0, \ldots, \mu_{n-1})$ with:

$$\mu_0 = 0$$

$$\begin{aligned} \mu_m &= 2k - 2\sum_{l=1}^{k} \cos\left(\frac{2\pi m l}{n}\right) \\ &= 2k + 1 - \frac{\sin\left(\left(k+\frac{1}{2}\right)2\pi\frac{m}{n}\right)}{\sin\left(\frac{\pi m}{n}\right)}, \quad m = 1, \ldots, n-1 \end{aligned}$$

where the second equality follows from the following trigonometric identity [Gradshteyn and Ryzhik (1994), page 36]:

$$\sum_{l=0}^{k} \cos(lx) = \frac{1}{2}\left(1 + \frac{\sin\left(\left(k+\frac{1}{2}\right)x\right)}{\sin\left(\frac{x}{2}\right)}\right)$$

A series expansion shows that as $n \to \infty$, $\lambda_2 \leq \mu_1 \approx \frac{4\pi^2(k+\frac{1}{2})^3}{3n^2}$. Since $\lambda_{\max} \geq 2k\frac{n}{n-1}$ by Lemma 7.11, $r \leq \frac{\mu_1}{\lambda_{\max}} \approx \frac{2\pi^2(k+\frac{1}{2})^3}{3kn^2}$. This means the values λ_2 and r decrease as $\Omega\left(\frac{1}{n^{1+3\epsilon}}\right)$ and $\Omega\left(\frac{1}{n^{\frac{4}{3}+2\epsilon}}\right)$ respectively. In particular, if k is bounded, i.e. $\epsilon = \frac{1}{3}$, then λ_2, r decrease as $\Omega\left(\frac{1}{n^2}\right)$ which is asymptotically the fastest possible by Lemma 7.11.

In the second case, we consider graphs for which $k < \lfloor\frac{n}{2}\rfloor$ and use the same approach as in [Atay *et al.* (2006)]. The main difference is that in [Atay *et al.* (2006)] the smallest nonzero eigenvalue of the *normalized* Laplacian matrix $\tilde{L} = I - D^{-1}A$ is studied[7].

Lemma 3.9 *[Atay* et al. *(2006)] Let $\mathcal{G}_1$ and $\mathcal{G}_2$ be two disjoint connected graphs with degree sequence $c = (c_1, \ldots, c_n)$ and $d = (d_1, \ldots, d_m)$ respectively. Then there exists a connected graph $\mathcal{H}$ with degree sequence $c \cup d$ and a subset of vertices S such that $e(S, \overline{S}) = 2$.*

Proof: Let (v_1, v_2) be an edge in $\mathcal{G}_1$ and (w_1, w_2) be an edge in $\mathcal{G}_2$. Consider the disjoint union of $\mathcal{G}_1$ and $\mathcal{G}_2$ and replacing the above 2 edges with the edges (v_1, w_1) and (v_2, w_2) and the results follows. This is schematically shown in Figure 3.5. □

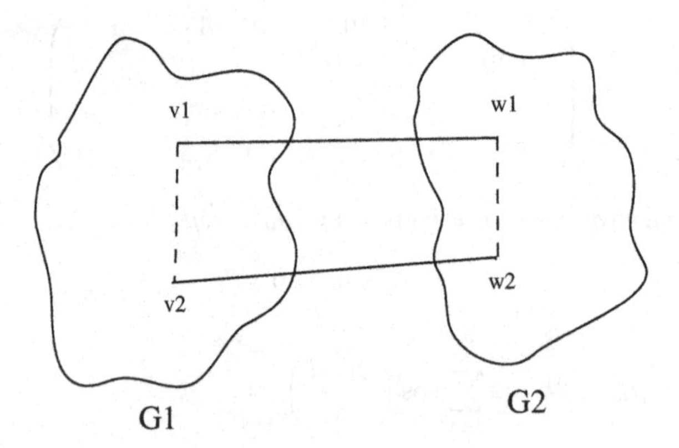

Fig. 3.5 Construction in [Atay *et al.* (2006)] to obtain a graph with $e(S, \overline{S}) = 2$.

As in [Atay *et al.* (2006)], we create a connected k-regular graph with $|S| = \lfloor\frac{n}{2}\rfloor$ and $e(S, \overline{S}) = 2$ where $e(\cdot, \cdot)$ is as defined in Definition A.1. By

[7] For k-regular graphs, this difference between L and $\tilde{L}$ is not important since $L = k\tilde{L}$ and the eigenvalues of L and $\tilde{L}$ differ by a constant factor k. However, for non-regular graphs, the eigenvalues of $\tilde{L}$ and L will have different properties.

Corollary A.3 this means that

$$\lambda_2(L) \leq \frac{2n}{\lfloor \frac{n}{2} \rfloor \lceil \frac{n}{2} \rceil} = \begin{cases} \frac{8}{n} & ,n \quad \text{even} \\ \frac{8}{n-\frac{1}{n}} & ,n \quad \text{odd} \end{cases} \leq \frac{8}{n-\frac{1}{n}}$$

$$r(L) \leq \begin{cases} \frac{8(n-1)}{kn^2} & ,n \quad \text{even} \\ \frac{8}{k(n+1)} & ,n \quad \text{odd} \end{cases} \leq \frac{8}{k(n+1)}$$

The reason for studying the two cases is that C_{2k} gives better bounds when k grows slower than $n^{\frac{1}{3}}$ and the construction in [Atay *et al.* (2006)] gives better bounds when k grows faster than $n^{\frac{1}{3}}$. In either case, $\lambda_2, r \to 0$ as $n \to \infty$.

3.7.1.2 *Construction 2: graph with high λ_2 and r*

With high probability, a random $2k$-regular graph has eigenvalues $\lambda_2 = 2k - O(\sqrt{k})$, $\lambda_{\max} = 2k + O(\sqrt{k})$ as $n \to \infty$ (Theorem 3.1). This means that on a relative scale, $\lambda_2 \approx 2k$, and $r \approx 1$ as $k, n \to \infty$ which can be considered optimal according to Corollary 2.37. In conclusion, whereas λ_2 and $r \to 0$ as $n \to \infty$ for Construction 1, they remain bounded from below for Construction 2.

Lemma 3.10 *If $\mathcal{H}$ is a subgraph of $\mathcal{G}$ with the same set of vertices*[8], *then*

(1) $\lambda_2(\mathcal{H}) + \lambda_2(\mathcal{G}\backslash\mathcal{H}) \leq \lambda_2(\mathcal{G})$,
(2) $\lambda_{\max}(\mathcal{H}) + \lambda_{\max}(\mathcal{G}\backslash\mathcal{H}) \geq \lambda_{\max}(\mathcal{G})$
(3) $\min(r(\mathcal{H}), r(\mathcal{G}\backslash\mathcal{H})) \leq \frac{\lambda_2(\mathcal{H})+\lambda_2(\mathcal{G}\backslash\mathcal{H})}{\lambda_{\max}(\mathcal{H})+\lambda_{\max}(\mathcal{G}\backslash\mathcal{H})} \leq r(\mathcal{G})$,
(4) $\lambda_{\max}(\mathcal{H}) \leq \lambda_{\max}(\mathcal{G})$.

Here $\mathcal{G}\backslash\mathcal{H}$ is a graph with the same set of vertices as $\mathcal{G}$ and with edges which are in $\mathcal{G}$ but not in $\mathcal{H}$. We use $\lambda_i(\mathcal{H})$ to denote λ_i of the Laplacian matrix of the graph $\mathcal{H}$.

Proof: Follows from the facts that $\lambda_2(L) = \min_{\sum_i x_i = 0, \|x\|=1} x^T Lx$, $\lambda_{\max}(L) = \max_{\|x\|=1} x^T Lx$ and $L(\mathcal{H}) + L(\mathcal{G}\backslash\mathcal{H}) = L(\mathcal{G})$. □

3.7.2 *Graphs with prescribed degree sequence*

For each n, consider a graphical list of degrees $0 < d_1 \leq d_2 \leq \cdots \leq d_n$. The average vertex degree is defined as $\overline{k} = \frac{\sum_i d_i}{n}$. Next we construct two

[8]By this we mean that $A_{ij} \leq B_{ij}$ for all i,j, where A and B are the adjacency matrices of $\mathcal{H}$ and $\mathcal{G}$ respectively.

connected graphs with the same list of degrees but different λ_2 and r.

3.7.2.1 *Construction 1: graph with low λ_2 and r*

As in Section 3.7.1.1 we consider two cases. In the first case, d_n grows as $\Omega\left(n^{\frac{1}{3}-\epsilon}\right)$ for $\epsilon > 0$. Given a graphical list of degrees $0 < d_1 \leq d_2 \leq \cdots \leq d_n$, the Havel-Hakimi algorithm in Theorem 3.7 constructs a connected graph with this degree sequence where the vertices are arranged on a line and each vertex is connected to its nearest neighbors. Let us denote this graph as $\mathcal{G}_{hh}$ with Laplacian matrix L_{hh}. The graph $\mathcal{G}_{hh}$ is a subgraph of the $2d_n$-regular graph C_{2d_n} defined in Sec. 3.7.1.1. Therefore, by Lemma 3.10, $\lambda_2(L_{hh}) \leq \mu_1 \approx \frac{4\pi^2(d_n+\frac{1}{2})^3}{3n^2} = \Omega\left(\frac{1}{n^{1+3\epsilon}}\right)$ as $n \to \infty$. Since $\lambda_{\max}(L_{hh}) \geq \frac{nd_n}{n-1}$, this means that $r(L_{hh}) \leq \frac{\mu_1}{\lambda_{\max}} \approx \frac{4\pi^2(d_n+\frac{1}{2})^3}{3d_n n^2} = \Omega\left(\frac{1}{n^{\frac{4}{3}+2\epsilon}}\right)$. As in Section 3.7.1.1, for bounded d_n the values of λ_2 and r for this construction decrease as $\Omega\left(\frac{1}{n^2}\right)$ when $n \to \infty$. By Lemma 7.11, this is the fastest possible rate.

In the second case, assume that the degree sequence is of the form $(d_{\max}, \ldots, d_{\max}, d_{\max}-1, \ldots, d_{\max}-1, \ldots, 1, \ldots, 1)$ with $d_{\max} \leq \frac{n}{4}$ and the average degree satisfies $\overline{k} = 2 + \frac{d_{\max}}{n}$. We further assume that for each $1 \leq i \leq d_{\max}$, there are at least 2 vertices with vertex degree i. As in [Atay *et al.* (2006)], we construct a graph with this degree sequence such that $|S| \geq \lceil \frac{n}{2} \rceil - d_{\max}$ and $e(S, \overline{S}) = 2$. By Corollary A.3 $\lambda_2(L) \leq \frac{e(S,\overline{S})n}{|S|(n-|S|)}$ and this means that

$$\lambda_2(L) \leq \frac{2n}{(\lfloor \frac{n}{2} \rfloor + d_{\max})(\lceil \frac{n}{2} \rceil - d_{\max})} = \begin{cases} \dfrac{8}{n - \frac{4d_{\max}^2}{n}} & , n \text{ even} \\[2ex] \dfrac{8}{n - \frac{4(d_{\max}-\frac{1}{2})^2}{n}} & , n \text{ odd} \end{cases}$$

$$\leq \frac{8}{n - \frac{4d_{\max}^2}{n}} \leq \frac{10\frac{2}{3}}{n}$$

$$r(L) \leq \frac{(n-1)\lambda_2(L)}{d_{\max} n} \leq \begin{cases} \dfrac{8(n-1)}{n^2 d_{\max} - 4d_{\max}^3} & , n \text{ even} \\[2ex] \dfrac{8(n-1)}{nd_{\max}\left(n - \frac{4(d_{\max}-\frac{1}{2})^2}{n}\right)} & , n \text{ odd} \end{cases}$$

$$\leq \frac{10\frac{2}{3}}{nd_{\max}}$$

where we have used the fact that $d_{\max} \leq \frac{n}{4}$. Again, in either case $r, \lambda_2 \to 0$ as $n \to \infty$.

3.7.2.2 *Construction 2: graph with high λ_2 and r*

For convenience, we allow loops in the graph, i.e. the graphs are not necessarily simple. Furthermore, rather than constructing graphs with a specific degree sequence, we consider a random graph model which has a prescribed degree sequence in expectation. Given a list of degrees $d_1 \leq d_2 \leq \cdots \leq d_n$ satisfying the condition

$$n(d_n - d_1)^2 \leq (n - d_1) \sum_k (d_k - d_1)$$

we construct a random graph as in Section 3.6 where an edge between vertex i and vertex j is randomly selected with probability

$$p_{ij} = \frac{d_1}{n} + \frac{(d_i - d_1)(d_j - d_1)}{\sum_k (d_k - d_1)}$$

Let us call this graph model $\mathcal{G}_{pr}$ with Laplacian matrix L_{pr}.

Assume now that $d_1 = p_1 n$ for some $0 < p_1 < 1$. By Theorem 3.3 and Lemma 3.8, $\lambda_2(\mathcal{G}_{pr}) \geq d_1 - o(\sqrt{n})$ with high probability. Since $\lambda_2 \leq \frac{n}{n-1} d_1$, this implies that $\lambda_2 \approx d_1$ and $r \gtrapprox \frac{d_1}{2d_n}$ for $\mathcal{G}_{pr}$, i.e. λ_2 and r are bounded away from 0 as $n \to \infty$.

The conclusion that $\lambda_2 \approx d_1$ shows that for the random graph model $\mathcal{G}_{pr}$, homogeneity of the degree sequence enhances synchronizability in terms of λ_2. In particular, for a given average degree $\overline{k}$, the highest λ_2 can be as $n \to \infty$ is approximately $\overline{k}$ since $\lambda_2 \leq \frac{n}{n-1} d_1 \leq \frac{n}{n-1} \overline{k}$. This is achieved for the random graph where $d_1 = \overline{k}$, i.e. the d_1-regular graph which has the most homogeneous degree sequence. This supports the experimental results in [Nishikawa *et al.* (2003)].

3.8 Further reading

The reader is referred to [Newman (2003); Bollobás and Riordan (2004)] for excellent surveys on the rapidly evolving research on complex networks and for more details on other characteristics of complex networks not discussed here, such as clustering coefficient, assortativity coefficient and degree correlation of a graph. See also the short article by Casselman [Casselman (2004)]. The book [Bollobás (2001)] is a classical introduction to the theory of random graphs.

Chapter 4

Synchronization in Networks of Nonlinear Continuous-time Dynamical Systems

In this chapter we study global synchronization phenomena in networks of identical systems, where each system is a continuous-time dynamical system. In particular, we study systems with the following state equation:

$$\dot{x} = \begin{pmatrix} f(x_1, t) \\ \vdots \\ f(x_n, t) \end{pmatrix} + (G(t) \otimes D(t))x + u(t) \tag{4.1}$$

where $x = \begin{pmatrix} x_1 \\ \vdots \\ x_n \end{pmatrix}$, $u = \begin{pmatrix} u_1 \\ \vdots \\ u_n \end{pmatrix}$ and $G(t)$ is a zero row sums matrix for all t. Each x_i is a state vector in $\mathbb{R}^m$. $G(t)$ is an n by n matrix and $D(t)$ is an m by m matrix for each t.

The matrix $G(t)$ describes the coupling topology whereas $D(t)$ is the linear coupling matrix between two systems. Since $G(t)$ is a zero row sum matrix for each time t, if all off-diagonal elements of $G(t)$ are nonpositive, then $G(t)$ is equal to the Laplacian matrix of the graph of the coupled system at time t. An example of a network of circuits which has the state equations in Eq. (4.1) is an array of Chua's circuits which are coupled via resistors [Wu and Chua (1995); Wu (1998b)].

If $D(t) \preceq 0$ and $G_{ij}(t) \leq 0$ for $i \neq j$ we call the coupling *cooperative*. If $D(t) \preceq 0$ and $G_{ij}(t) \geq 0$ for $i \neq j$ we call the coupling *competitive*. One of the conclusions of this chapter is that under certain conditions, strong cooperative coupling will result in a synchronized network.

Definition 4.1 Given a matrix V, a function $f(y, t)$ is V-uniformly decreasing if $(y - z)^T V(f(y, t) - f(z, t)) \leq -c\|y - z\|^2$ for some $c > 0$ and all y, z, t.

The condition of V-uniformly decreasing for a real matrix $V = V^T \succ 0$ implies that $\dot{y} = f(y,t)$ is globally asymptotically stable. In particular, if $f(y,t) + K(t)y$ is V-uniformly decreasing for some real matrix $V = V^T \succ 0$, then the term $K(t)x$ is a linear state feedback that globally stabilizes the system $\dot{x} = f(x,t)$. This can be seen by using $x^T V x$ as the quadratic Lyapunov function. This property is also known as quadratically stabilizable in the control systems literature. For instance, the classical Chua's circuit can be quadratically stabilized by choosing $V = I$ and $K(t) = -\kappa\ \mathrm{diag}(1,0,0)$ for a large enough scalar $\kappa > 0$ [Wu and Chua (1994)]. For differentiable f, the Mean Value Theorem shows that the condition above is equivalent to $V\left(\frac{\partial f(x,t)}{\partial x} + K(t)\right) + \delta I$ being negative definite for some $\delta > 0$ [Chua and Green (1976)].

Definition 4.2 $\mathcal{W}$ is the set of square real matrices with zero row sums and nonpositive off-diagonal elements. $\mathcal{W}_s$ is the set of irreducible symmetric matrices in $\mathcal{W}$.

Note that Laplacian matrices of graphs belong to the set $\mathcal{W}$ and Laplacian matrices of connected undirected graphs are in the set $\mathcal{W}_s$.

Definition 4.3 $\mathcal{M}$ is the synchronization manifold defined as the linear subspace $\{x : x_i = x_j, \forall i,j\}$. If x approaches the set $\mathcal{M}$, i.e. $\|x_i - x_j\| \to 0$ as $t \to \infty$ for all i, j, then the coupled network in Eq. (4.1) is said to *synchronize*.

An element of $\mathcal{M}$ can be written as $\mathbf{1} \otimes z$. Consider the following synchronization result in [Wu and Chua (1995); Wu (2003a, 2005g)]:

Theorem 4.4 *Let $Y(t)$ be an m by m time-varying matrix and V be an m by m symmetric positive definite matrix such that $f(x,t) + Y(t)x$ is V-uniformly decreasing. Then the network of coupled dynamical systems in Eq. (4.1) synchronizes if the following two conditions are satisfied:*

(1) $\lim_{t\to\infty} \|u_i - u_j\| = 0$ for all i, j,
(2) There exists an n by n matrix $U \in \mathcal{W}_s$ such that

$$(U \otimes V)(G(t) \otimes D(t) - I \otimes Y(t)) \preceq 0$$

for all t.

Proof: We prove this via Lyapunov's second method. Construct the Lyapunov function $g(x) = \frac{1}{2}x^T(U \otimes V)x$. By Lemma 2.18, the hypotheses imply that the derivative of g along trajectories of Eq. (4.1) is

$$\begin{aligned}
\dot{g} &= x^T(U \otimes V)\dot{x} \\
&= x^T(U \otimes V)\begin{pmatrix} f(x_1,t) + Y(t)x_1 + u_1(t) \\ \vdots \\ f(x_n,t) + Y(t)x_n + u_n(t) \end{pmatrix} \\
&\quad + x^T(U \otimes V)(C(t) \otimes D(t) - I \otimes Y(t))x \\
&\leq \textstyle\sum_{i<j} -U_{ij}(x_i - x_j)^T V(f(x_i,t) + Y(t)x_i - f(x_j,t) - Y(t)x_j) \\
&\quad + \textstyle\sum_{i<j} -U_{ij}(x_i - x_j)^T V(u_i(t) - u_j(t)) \\
&\leq \textstyle\sum_{i<j} -U_{ij}(-\mu\|x_i - x_j\|^2 + (x_i - x_j)^T V(u_i(t) - u_j(t)))
\end{aligned}$$

Note that $-U_{ij} \geq 0$ for $i < j$. For each $-U_{ij} > 0$ and $\delta > 0$, and sufficiently large t, $(u_i(t) - u_j(t))$ is small enough such that if $\|x_i - x_j\| \geq \delta$, then $\dot{g} \leq -\frac{\mu}{2}\|x_i - x_j\|^2$. This implies that for large enough t, $\|x_i - x_j\| < \delta$. Therefore $\lim_{t\to\infty} \|x_i - x_j\| = 0$. Irreducibility of U implies that enough U_{ij} are nonzero to ensure $\|x_i - x_j\| \to 0$ for all i and j (see Lemma 2.18). □

An alternative way to write Eq. (4.1) is

$$\dot{x} = \begin{pmatrix} f_1(x_1,t) \\ \vdots \\ f_n(x_n,t) \end{pmatrix} + (G(t) \otimes D(t))x$$

where $f_i(x_i,t) = f(x_i,t) + u_i(t)$.

4.1 Static coupling topology

In this case, the matrix G in Eq. (4.1) is assumed to be a constant matrix, although $D(t)$ can still change with time. To characterize how amenable the coupling topology, expressed as G, is to synchronization, we introduce the following quantity $\mu(G)$:

Definition 4.5 For a real matrix G with nonpositive off-diagonal elements, $\mu(G)$ is defined as the supremum of the set of real numbers μ such that $U(G - \mu I) \succeq 0$ for some $U \in \mathcal{W}_s$.

We have the following Corollary to Theorem 4.4:

Corollary 4.6 *The network of systems described by the state equations*

$$\dot{x} = \begin{pmatrix} f(x_1, t) \\ \vdots \\ f(x_n, t) \end{pmatrix} + (G \otimes D(t))x + u(t) \tag{4.2}$$

synchronizes if

(1) $\lim_{t\to\infty} \|u_i - u_j\| = 0$ *for all* i, j,
(2) $VD(t) = D^T(t)V \preceq 0$ *for all* t,
(3) There exists a symmetric positive definite matrix V *such that* $f(y,t) + \mu(G)D(t)y$ *is* V*-uniformly decreasing.*

Proof: Apply Theorem 4.4 with $Y(t) = \mu(G)D(t)$ we get the condition

$$(U \otimes V)((G - \mu(G)I) \otimes D(t)) \preceq 0$$

which is equivalent to $U(G - \mu(G)I) \succeq 0$ since $VD(t) = D^T(t)V \preceq 0$. This condition is satisfied by the definition of $\mu(G)$. □

The condition $VD(t) = D^T(t)V \preceq 0$ is satisfied in several situations studied in the literature. Among others it is satisfied if one of the following conditions is satisfied:

(1) V and $D(t)$ are both diagonal and $D \preceq 0$.
(2) $D(t)$ is a nonpositive multiple of the identity matrix.
(3) V is a positive multiple of the identity matrix and $D(t) = D^T(t) \preceq 0$.

The larger $\mu(G)$ is, the less the coupling $D(t)$ needs to be to satisfy condition (3) in Corollary 4.6. Thus Corollary 4.6 suggests that networks for which $\mu(G)$ is large are easier to synchronize.

4.1.1 *Properties of* $\mu(G)$

Since matrices in $\mathcal{W}_s$ are positive semidefinite, the set of real numbers μ such that $U(G-\mu I) \succeq 0$ for some $U \in \mathcal{W}_s$ is an interval, i.e. if $U(G-\mu I) \succeq 0$ for some $U \in \mathcal{W}_s$, then $U(G - \lambda I) \succeq 0$ for all $\lambda \leq \mu$.

Lemma 4.7 *[Wu and Chua (1995)] Let* X *be a real matrix. If* $A \in \mathcal{W}_s$ *and either* $AX \succeq 0$ *or* $AX \preceq 0$, *then* X *is a matrix with constant row sums.*

Proof: First note that $\mathbf{1}^T(AX + X^TA)\mathbf{1} = 0$. Suppose $AX \succeq 0$. Express $\mathbf{1}$ as a linear combination of eigenvectors of $AX + X^TA$, i.e. $\mathbf{1} = \sum_i \alpha_i v_i$.

Since $AX + X^T A$ is symmetric positive semidefinite, all eigenvectors are orthogonal and all eigenvalues are nonnegative. Since $\mathbf{1}^T(AX + X^T A)\mathbf{1} = \sum_i \lambda_i \alpha_i^2 \|v_i\|^2 = 0$, this implies that $\mathbf{1}$ is a linear combination of eigenvectors corresponding to the zero eigenvalue, i.e. $\lambda_i = 0$ and $\mathbf{1}$ is a eigenvector of $AX + X^T A$ with eigenvalue 0. Thus $(AX + X^T A)\mathbf{1} = AX\mathbf{1} = \mathbf{0}$. By Theorem 2.7, the zero eigenvalue of A is simple, and this implies that $X\mathbf{1} = \epsilon\mathbf{1}$ for some ϵ which implies that that X is a matrix with constant row sums. The case of $AX \preceq 0$ is similar. □

Lemma 4.7 implies that $\mu(G)$ is only defined when the matrix G has constant row sums. A matrix with constant row sums can be converted to a matrix with zero row sums by adding a multiple of the identity matrix. Thus for the purpose of finding $\mu(G)$ we can assume without loss of generality that G has zero row sums. In other words, adding αI to G shifts $\mu(G)$ by α. For a matrix with zero row sums, 0 is an eigenvalue with eigenvector $\mathbf{1}$. The next theorem shows that the quantity $\mu(G)$ exists for zero row sum matrices and gives a lower bound.

Theorem 4.8 *If G has zero row sums, then $\mu(G)$ exists, i.e. there is a real number μ and a matrix $U \in \mathcal{W}_s$ such that $U(G-\mu I) \preceq 0$. Furthermore,*

$$\mu(G) \geq a_1(G) \geq \lambda_{\min}\left(\frac{1}{2}(G + G^T)\right)$$

where a_1 is the algebraic connectivity defined in Sect. 2.3.2.

Proof: Let J be the n by n matrix of all 1's and let $M = I - \frac{1}{n}J$. It is clear that $M \in \mathcal{W}_s$. Let $U = M$. Define the symmetric matrix $H = \frac{1}{2}(U(G - \mu I) + (U(G - \mu I))^T) = \frac{1}{2}(G + G^T) - \mu M - \frac{1}{2n}(JG + G^T J)$. Since $J\mathbf{1} = n\mathbf{1}$ and $G\mathbf{1} = M\mathbf{1} = 0$, it follows that $H\mathbf{1} = 0$. Let $x \perp \mathbf{1}$ with $\|x\| = 1$. This means that $Mx = x$. Then $x^T H x = \frac{1}{2}x^T(G + G^T)x - \mu - \frac{1}{2n}x^T(JG + G^T J)x$. Since $x \perp \mathbf{1}$, this implies $Jx = 0$ and thus $x^T H x = \frac{1}{2}x^T(G + G^T)x - \mu$ which means that $H \succeq 0$ if $\mu \leq \frac{1}{2}x^T(G + G^T)x$. This implies $\mu(G) \geq a_1(G)$. The proof is then complete by noting that $a_1(G) \geq \min_{\|x\|=1} x^T G x = \lambda_{\min}\left(\frac{1}{2}(G + G^T)\right)$. □

Definition 4.9 For a matrix G with zero row sums, let $\mathcal{L}(G)$ denote the eigenvalues of G that do not correspond to the eigenvector $\mathbf{1}$.

Corollary 4.10 *If G is a real matrix with zero row sums and zero column sums, then $\mu(G) \geq \lambda_2^s(G)$ where $\lambda_2^s(G)$ is the smallest eigenvalue in $\mathcal{L}(\frac{1}{2}(G + G^T))$.*

Proof: Since $\mathbf{1}$ is an eigenvector of $\frac{1}{2}(G+G^T)$, we have $\lambda_2^s(G) = a_1(G)$. The result then follows from Theorem 4.8. □

Corollary 4.11 *If $G \in \mathcal{W}$, then $\mu(G) \geq a_1(G) \geq 0$. If in addition $G + G^T$ is irreducible, then $\mu(G) \geq a_1(G) > 0$.*

Proof: For a symmetric matrix $X \in \mathcal{W}$, $\lambda_2(X) \geq 0$. For a matrix $X \in \mathcal{W}_s$, $\lambda_2(X) > 0$. This is a consequence of Theorem 2.7. The theorem then follows from Theorem 4.8 and the fact that $G + G^T \in \mathcal{W}$. □

Definition 4.12 For a real matrix G with zero row sums, define $\mu_2(G)$ as $\mu_2(G) = min_{\lambda \in \mathcal{L}(G)} \mathcal{R}e(\lambda)$ where $\mathcal{R}e(\lambda)$ is the real part of λ.

Next we show an upper bound for $\mu(G)$.

Theorem 4.13 *If G is a real matrix with zero row sums, then $\mu(G) \leq \mu_2(G)$.*

Proof: Let $\lambda \in \mathcal{L}(G)$ with corresponding eigenvector v. Let $U \in \mathcal{W}_s$ be such that $U(G - \mu I) \succeq 0$ for some real number μ. The kernel of U is spanned by $\mathbf{1}$. By definition of $\mathcal{L}(G)$, v is not in the kernel of U. Since $(G - \mu I)v = (\lambda - \mu)v$, this implies that $v^*U(G - \mu I)v = (\lambda - \mu)v^*Uv$. Positive semidefiniteness of $U(G - \mu I)$ implies that $\mathcal{R}e(v^*U(G - \mu I)v) \geq 0$. Since U is symmetric positive semidefinite and v is not in the kernel of U, $v^*Uv > 0$. This implies that $\mathcal{R}e(\lambda) - \mu \geq 0$. □

Corollary 4.14 *If G is a real matrix with zero row sums, then*

$$\lambda_{\min}\left(\frac{1}{2}(G + G^T)\right) \leq a_1(G) \leq \mu_2(G).$$

Proof: Follows from Theorems 4.8 and 4.13. □

Theorems 4.8 and 4.13 show that $a_1(G) \leq \mu(G) \leq \mu_2(G)$. Next we present two classes of matrices for which there is a closed form expression for $\mu(G)$.

Theorem 4.15 *If G is a real normal matrix with zero row sums, then $a_1(G) = \mu(G) = \mu_2(G)$.*

Proof: Since G is a normal matrix, an eigenvector of G with eigenvalue δ is an eigenvector of G^T with eigenvalue $\overline{\delta}$ [Gantmacher (1960)]. Applying this to the eigenvector $\mathbf{1}$ shows that $G^T\mathbf{1} = 0$, i.e. G has zero column sums. Furthermore, for a real normal matrix, the eigenvalues of $\frac{1}{2}(G + G^T)$ are just the real parts of the eigenvalues of G [Horn and Johnson (1985)]. This implies that $\mu_2(G) = \lambda_2^s(G)$. The result then follows from Corollary 4.10 and Theorem 4.13. $\square$

Lemma 4.16 *[Wu and Chua (1995)] For any n by n constant row sum matrix A, there exists $n-1$ by $n-1$ matrix B such that $CA = BC$ where C is defined as*

$$C = \begin{pmatrix} 1 & -1 & & & \\ & 1 & -1 & & \\ & & & \cdots & \\ & & & & 1 & -1 \end{pmatrix} \tag{4.3}$$

In particular, $B = CAR$ where

$$R = \begin{pmatrix} 1 & 1 & 1 & \dots & 1 \\ 0 & 1 & 1 & \dots & 1 \\ & & \ddots & & 1 \\ & & & 1 & 1 \\ 0 & 0 & \dots & 0 & 1 \\ 0 & 0 & \dots & 0 & 0 \end{pmatrix}$$

We denote this as $B = \Upsilon(A)$. Furthermore, $\Upsilon(A + \alpha I) = \Upsilon(A) + \alpha I$ and Spectrum$(A) = \{$Spectrum$(B), \epsilon\}$ where ϵ is the row sum of A.

Proof: First note that CA is a zero row sums matrix. The n by n matrix RC is

$$RC = \begin{pmatrix} 1 & 0 & \cdots & 0 & -1 \\ 0 & 1 & 0 & \vdots & -1 \\ 0 & 0 & \ddots & 0 & -1 \\ 0 & 0 & \cdots & 1 & -1 \\ 0 & 0 & \cdots & 0 & 0 \end{pmatrix}$$

This means that the first $n-1$ columns of $CARC$ is the same as those of CA. The n-th column of $CARC$ is the negative of the sum of the first $n-1$ columns of CA which is equal to the n-th column of CA since CA is a zero row sums matrix. Thus $CARC = CA$. Since CR is the $n-1$ by $n-1$ identity matrix, we have $\Upsilon(A + \alpha I) = \Upsilon(A) + \alpha I$. Since

$$\begin{pmatrix} & C & \\ 0 & \cdots & 1 \end{pmatrix} (\lambda I - A) \begin{pmatrix} & 1 \\ R & \vdots \\ & 1 \end{pmatrix} = \begin{pmatrix} \lambda I - B & 0 \\ & \vdots \\ w & \lambda - \epsilon \end{pmatrix}$$

where w is the last row of $-AR$, the spectrum of A is the spectrum of B plus ϵ. □

Theorem 4.17 *[Wu (2003b)] If G is a triangular zero row sums matrix, then $\mu(G) = \mu_2(G)$. For instance, if G is an upper triangular zero row sums matrix, then $\mu(G)$ is equal to the least diagonal element of G, excluding the lower-right diagonal element.*

Proof: Without loss of generality, suppose that G is an $n \times n$ upper triangular zero sum matrix:

$$G = \begin{pmatrix} a_{1,1} & a_{1,2} & \cdots & & a_{1,n} \\ 0 & a_{2,2} & a_{2,3} & \cdots & a_{2,n} \\ 0 & 0 & \ddots & & \\ 0 & 0 & 0 & a_{n-1,n-1} & -a_{n-1,n-1} \\ 0 & 0 & \cdots & & 0 \end{pmatrix} \tag{4.4}$$

From Theorem 4.13, $\mu(G) \leq \min_{1 \leq i \leq n-1} a_{i,i}$. By Lemma 4.16,

$$B = \begin{pmatrix} a_{1,1} & b_{1,2} & \cdots & \\ & a_{2,2} & b_{2,3} & \cdots \\ & & \ddots & \\ & & & a_{n-1,n-1} \end{pmatrix}$$

is an $(n-1) \times (n-1)$ upper triangular matrix which satisfies $CG = BC$ where C is defined in Eq. (4.3). Let $\Delta = \text{diag}(\alpha_1, \ldots \alpha_{n-1})$ where $\alpha_i > 0$ and $H = \Delta B \Delta^{-1}$. The (i,j)-th element of H is $b_{i,j}\frac{\alpha_i}{\alpha_j}$ if $j > i$ and 0 if $j < i$. Therefore, for each $\epsilon > 0$, if we choose α_j much larger than α_i for all $j > i$, then we can ensure that the (i,j)-th element of H has absolute value less than $\frac{2\epsilon}{n-2}$ for $j > i$. By Theorem 2.9 the eigenvalues of $\frac{1}{2}(H + H^T)$ are larger than $\min_{1 \leq i \leq n-1} a_{i,i} - \epsilon$. Consider $U = C^T \Delta^2 C$ which is a matrix in $\mathcal{W}_s$ by Lemma 2.17. The matrix

$$U(G - \mu I) = C^T \Delta^2 C(G - \mu I) = C^T \Delta (H - \mu I) \Delta C$$

is positive semidefinite if $H - \mu I$ is positive semidefinite. From the discussion above $H - \mu I \geq 0$ if $\mu \leq \min_{1 \leq i \leq n-1} a_{i,i} - \epsilon$. Since ϵ is arbitrary, $\mu(G) \geq \min_{1 \leq i \leq n-1} a_{i,i}$. □

Corollary 4.18 *If L is the Laplacian matrix of the reversal of a directed tree, then $\mu(L) = \mu_2(L) = 1$.*

Proof: First note that the Laplacian matrix of an acyclic digraph can always have its rows and columns be simultaneously permuted to an upper-triangular matrix, say $\tilde{L}$. Since $\tilde{L}$ has zero row sums, $\tilde{L}_{nn} = 0$ and thus $\mu_2(L) = \mu_2(\tilde{L}) = \min_{i<n} \tilde{L}_{ii}$. Thus for an acyclic graph $\mu_2(L) = \min_{i \neq j} L_{ii}$ where j is an index such that $L_{jj} = 0$. Since L_{ii} are the indegrees of the interaction graph, this in particular implies that $\mu_2(L) = 1$ if the interaction graph of L is a directed tree. □

4.1.2 *Computing* $\mu(G)$

A semidefinite programming problem (SDP) is a convex nonlinear programming problem with a linear cost function and constraints which are linear or semidefinite. In general, a semidefinite programming problem can be written as [Vandenberghe and Boyd (1996)]:

$$\min_x c^T x \quad \text{subject to} \quad B_0 + \sum_i x_i B_i \succeq 0$$

In a feasibility SDP problem, the cost function is omitted; the goal is to determine whether there exists x that satisfies the constraints. SDP problems can be solved efficiently using interior point methods [Nesterov and Nemirovskii (1994)] and several software packages, both commercial and in the public domain, have been developed to solve SDP problems. The reader is referred to `http://www-user.tu-chemnitz.de/~helmberg/sdp_software.html` for a list of available software for solving SDP problems.

The value of $\mu(G)$ can be computed by solving a series of SDP problems [Wu (2006a)]. For a fixed μ, finding a matrix U such that $U(G - \mu I) \succeq 0$ is an SDP problem. Clearly $U(G - \mu I) \succeq 0$ is a linear matrix inequality. The property that U is symmetric, have zero row sums and with nonpositive off-diagonal elements can also be cast as matrix constraints. As for the irreducibility of U, this corresponds to the zero eigenvalue of U being simple and this is equivalent to showing $Q^T U Q \succ 0$ where Q is defined as in Theorem 2.35(1). To ensure that U will not be too small, we will use the constraint $Q^T U Q \succeq I$ instead. This does not affect the value of $\mu(G)$ since $U(G - \mu I) \succeq 0$ if and only if $\alpha U(G - \mu I) \succeq 0$ for any scalar $\alpha > 0$.

Thus we arrive at the following feasibility SDP problem:

$$\text{Find } U = U^T \text{ such that} \tag{4.5}$$
$$U(G - \mu I) \succeq 0,$$
$$U\mathbf{1} = 0,$$
$$U_{ij} \leq 0 \text{ for all } i \neq j,$$
$$Q^T U Q \succeq I.$$

Note that by Theorems 4.8 and 4.13, $\mu(G)$ can be bounded in the interval $[a_1, \mu_2]$. Since the set of values of μ such that $U(G - \mu I) \succeq 0$ for some $U \in \mathcal{W}_s$ is an interval, we can compute $\mu(G)$ by using the bisection method to successively refine μ and then solving the corresponding SDP problem (4.5). This is shown in Algorithm 1 where the value of ub is initially set to $\mu_2(G)$ and the value of lb is initially set to $a_1(G)$.

Algorithm 1 Compute $\mu(G)$

```
μ ← ub
if Problem (4.5) is infeasible then
  while |ub − lb| > ε do
    μ ← ½(ub + lb)
    if Problem (4.5) is infeasible then
      ub ← μ
    else
      lb ← μ
    end if
  end while
end if
μ(G) ← μ
```

Next we present computer simulation results obtained using CSDP 4.7 [Borchers (1999)] with the YALMIP 3 MATLAB interface (`http://control.ee.ethz.ch/~joloef/yalmip.msql`) to solve the SDP problem.

4.1.3 *Zero row sums matrices*

Our computer results are summarized in Table 4.1. Zero row sums matrices of small order are generated and their values of $\mu(G)$ are computed. For each order n, 5000 zero row sums matrices are chosen by generating the off-diagonal elements independently from a uniform distribution in the interval $\left[-\frac{1}{2}, \frac{1}{2}\right]$. The matrices are categorized into two groups depending on whether all their eigenvalues are real or not. For each group, the mean and the standard deviation of the quantities $0 \le i(G) = \frac{\mu(G)-a_1(G)}{\mu_2(G)-a_1(G)} \le 1$ and $r(G) = \frac{\mu(G)}{\mu_2(G)}$ are listed.

	Only real eigenvalues				Real and complex eigenvalues			
	$i(G)$		$r(G)$		$i(G)$		$r(G)$	
Order	mean	std	mean	std	mean	std	mean	std
3	0.9862	0.0621	1.0166	1.2065	0.8672	0.1724	2.1340	34.0588
4	0.9766	0.0777	1.0227	0.9181	0.5251	0.4002	0.2362	56.8556
5	0.9715	0.0809	1.0113	0.0589	0.3191	0.3353	3.0385	133.0064
10	0.9731	0.0894	1.0060	0.0214	0.1451	0.1095	-2.8780	424.5138

Table 4.1: Statistics of $i(G) = \frac{\mu(G)-a_1(G)}{\mu_2(G)-a_1(G)}$ and $r(G) = \frac{\mu(G)}{\mu_2(G)}$ for zero row sum matrices.

We observe significant differences between the behavior of $i(G)$ and $r(G)$ for matrices with only real eigenvalues and for matrices with complex eigenvalues. In particular, we see that $i(G)$ is close to 1 for matrices with only real eigenvalues which implies that $\mu(G)$ is close to $\mu_2(G)$ in this case. On the other hand, for matrices with complex eigenvalues, the statistics of $i(G)$ show that $\mu(G)$ is usually significantly less than $\mu_2(G)$.

4.1.4 *Matrices in $\mathcal{W}$*

Our computer results for matrices in $\mathcal{W}$ are summarized in Table 4.2. For each order n, 5000 zero row sums matrices are chosen by generating the off-diagonal elements independently from a uniform distribution in the interval $[-1, 0]$. The matrices are categorized into two groups depending on whether all their eigenvalues are real or not. For each group, the mean and the standard deviation of the quantities $i(G)$ and $r(G)$ are listed.

	Only real eigenvalues				Real and complex eigenvalues			
	$i(G)$		$r(G)$		$i(G)$		$r(G)$	
Order	mean	std	mean	std	mean	std	mean	std
3	0.9862	0.0612	0.9984	0.0084	0.8663	0.1698	0.9777	0.0337
4	0.9816	0.0658	0.9977	0.0095	0.9196	0.1389	0.9875	0.0246
5	0.9758	0.0770	0.9971	0.0112	0.9307	0.1292	0.9901	0.0211
10	0.9818	0.0260	0.9987	0.0020	0.9333	0.1097	0.9936	0.0123

Table 4.2: Statistics of $i(G) = \frac{\mu(G)-a_1(G)}{\mu_2(G)-a_1(G)}$ and $r(G) = \frac{\mu(G)}{\mu_2(G)}$ for matrices in $\mathcal{W}$.

In contrast to general zero row sums matrices, the behaviors of $\mu(G)$ for matrices in $\mathcal{W}$ with only real eigenvalues and for matrices in $\mathcal{W}$ with complex eigenvalues are similar. Furthermore, $\mu(G)$ is very close to $\mu_2(G)$, especially for matrices with only real eigenvalues.

4.1.5 *Synchronization and algebraic connectivity*

Theorem 4.8 shows that a_1 is a lower bound for μ. The next result shows that other forms of algebraic connectivity (Section 2.3.2) are also lower bounds for $\mu(G)$.

Theorem 4.19 *If G is the Laplacian matrix of a strongly connected graph, then $0 < a_2(G) \leq a_3(G) \leq \mu(G)$.*

Proof: Let w be a nonnegative vector such that $w^T G = 0$ and $\|w\|_\infty = 1$. This vector exists by Theorem 2.7. Let W be a diagonal matrix with the elements of w on the diagonal. Define $U = W - \frac{ww^T}{\|w\|_1}$. Then $U \in \mathcal{W}_s$. Next we show that $B = U(G - a_3(G)I)$ is positive semidefinite. First note that since $\mathbf{1}^T U = U\mathbf{1} = 0$, $\mathbf{1}^T B = B\mathbf{1} = 0$. Thus it suffices to show that $\min_{y \perp \mathbf{1}} y^T B y \geq 0$. Since $w^T G = 0$, B can be written as $B = WG - a_3(G)U$. For $y \perp \mathbf{1}$, $y^T B y = y^T W L y - a_3(G) y^T U y \geq 0$, since $y^T W L y \geq a_3(G) y^T U y$ by definition. This shows that $a_3(G) \leq \mu(G)$. The proof is then complete via Theorem 2.47. □

Since a_i are variants of the concepts of algebraic connectivity, this indicates a relationship between synchronizability and algebraic connectivity, i.e. the higher the algebraic connectivity (indicating a more tightly connected coupling topology), the easier it is to synchronize the network.

4.2 Coupling topology with a spanning directed tree

In this section, we show the following intuitive result: if there is a system which influences directly or indirectly all other systems, then the network synchronizes under sufficiently large cooperative coupling.

Theorem 4.20 *Consider a network of dynamical systems with state equation Eq. (4.1). The network synchronizes if the following conditions are satisfied:*

(1) $G \in W$,
(2) $\forall i, j$, $\lim_{t\to\infty} \|u_i - u_j\| = 0$,

(3) *$f(x,t) + a_4(G)D(t)x$ is V-uniformly decreasing for some symmetric positive definite V,*

(4) *$VD(t) = D^T(t)V \preceq 0$ for all t,*

Proof: Without loss of generality, we can assume that that $P = I$ in the Frobenius normal form of G (Eq. (2.1)). Let $\tilde{x}$ and $\tilde{u}$ be the part of the state vector x and input vector u corresponding to B_k, with $\tilde{x} = \begin{pmatrix} x_s \\ x_{s+1} \\ \vdots \\ x_n \end{pmatrix}$, $\tilde{u} = \begin{pmatrix} u_s \\ u_{s+1} \\ \vdots \\ u_n \end{pmatrix}$. The state equation for $\tilde{x}$ is then

$$\dot{\tilde{x}} = \begin{pmatrix} f(x_s,t) \\ \vdots \\ f(x_n,t) \end{pmatrix} + (B_k \otimes D(t))\tilde{x} + \tilde{u} \tag{4.6}$$

By Theorem 4.4 the network in Eq. (4.6) synchronizes if there exists a matrix $\tilde{U} \in \mathcal{W}_s$ such that $(\tilde{U} \otimes V)((B_k - a_4(G)I) \otimes D(t)) \preceq 0$. Since $VD(t)$ is symmetric negative semidefinite, this is equivalent to $\tilde{U}(B_k - a_4(G)I) \succeq 0$. Since by definition $a_3(B_k) \geq a_4(G)$, we have by Theorem 4.19 $\mu(B_k) \geq a_4(G)$ and thus such a matrix $\tilde{U}$ exists and the network in Eq. (4.6) synchronizes and thus $\lim_{t\to\infty} \|x_i - x_j\| = 0$ for $s \leq i \leq j \leq n$. Since the systems in Eq. (4.6) are synchronized, we can collapse their dynamics to that of a single system. In particular, the state equation for x_s can be written as $\dot{x}_s = f(x_s,t) + u_s(t) + \phi_s(t)$ where $\phi_s(t) \to 0$ as $t \to \infty$.

Let us rewrite the Frobenius normal form of G as

$$G = \begin{pmatrix} F+C & H \\ & B_k \end{pmatrix}$$

where F is a square zero row sums matrix and C is diagonal. Note that $F + C$ is block upper-triangular with the diagonal blocks equal to $B_1, \ldots, B_{k-1}$. Then the dynamics of $(x_1, \ldots, x_s)$ can be written as:

$$\begin{pmatrix} \dot{x}_1 \\ \vdots \\ \dot{x}_s \end{pmatrix} = \begin{pmatrix} f(x_1,t) \\ \vdots \\ f(x_s,t) \end{pmatrix} + \left(\tilde{G} \otimes D(t)\right) \begin{pmatrix} x_1 \\ \vdots \\ x_s \end{pmatrix} + \begin{pmatrix} \phi_1(t) + u_1(t) \\ \vdots \\ \phi_s(t) + u_s(t) \end{pmatrix} \tag{4.7}$$

where

$$\tilde{G} = \begin{pmatrix} F + C & h \\ 0 & 0 \end{pmatrix}$$

and $\phi_i(t) \to 0$ as $t \to \infty$ and h is a vector of the row sums of H. Since G has zero row sums, this means that the elements of $-h$ are equal to the diagonal elements of C. The functions ϕ_i can be considered as residual errors that occurred when replacing x_i, $i > s$ in the state equation Eq. (4.1) with x_s. Construct the following matrices

$$R = \begin{pmatrix} I & -\mathbf{1} \end{pmatrix}, \quad Q = \begin{pmatrix} I \\ 0 \end{pmatrix}$$

If G is an n by n matrix and B_k is an l by l matrix, then the dimensions of $\tilde{G}$, R, Q and F are $n-l+1$ by $n-l+1$, $n-l$ by $n-l+1$, $n-l+1$ by $n-l$ and $n-l$ by $n-l$ respectively. It is easy to verify that $R\tilde{G}Q = F + C$ and $R\tilde{G}QR = R\tilde{G}$. Let $W = \mathrm{diag}(w_1, \ldots w_{k-1})$ where w_i are positive vectors such that $w_i^T L_i = 0$ and $\max_v w_i(v) = 1$. Note that $I \succeq W \succ 0$.

Let $\Delta = \mathrm{diag}(\alpha_1 I_1, \ldots, \alpha_k I_{k-1})$ where I_j are identity matrices of the same dimension as B_j and $0 < \alpha_j \leq 1$. Let $Z = \Delta R\tilde{G}Q\Delta^{-1}$. If we choose α_j much larger than α_i for $j > i$, then Z is nearly block-diagonal with the blocks equal to B_1, B_2, $\ldots B_{k-1}$. Now let $U = R^T \Delta W \Delta R$. It is easy to see that U is a matrix in $\mathcal{W}_s$.

Then $U(\tilde{G} - a_4(G)I) = R^T \Delta W \Delta R(\tilde{G} - a_4(G)I) = R^T \Delta(WZ - a_4(G)W)\Delta R$. By choosing appropriate α_j's Z can be made as close to block diagonal as possible. Since $\eta_i \geq a_4(G)$ for $i < k$ it follows that $WZ - a_4(G)W \succeq -\epsilon I$ for arbitrarily small ϵ and Theorem 4.4 can again be applied to show that Eq. (4.7) synchronizes[1]. □

Assuming the conditions in Theorem 4.20 are satisfied, the quantity $a_4(G)$, which depends on the underlying graph, provides a bound on the amount of coupling needed to synchronize the network.

Corollary 4.21 *Consider the network of coupled dynamical systems with state equations:*

$$\dot{x} = \begin{pmatrix} f(x_1, t) \\ \vdots \\ f(x_n, t) \end{pmatrix} + \kappa(G \otimes D(t))x + u(t) \tag{4.8}$$

where κ is a scalar. Assume that the following conditions are satisfied:

[1]Where η_i is defined as in Definition 2.32.

(1) $G \in W$,
(2) $\forall i, j$, $\lim_{t\to\infty} \|u_i - u_j\| = 0$,
(3) $f(x,t) + D(t)x$ *is* V*-uniformly decreasing for some symmetric positive definite* V,
(4) $VD(t) = D^T(t)V \preceq 0$ *for all* t,

If the interaction graph of G *contains a spanning directed tree, then the network in Eq. (4.8) synchronizes for sufficiently large* $\kappa > 0$.

Proof: By Theorem 4.20 the network synchronizes if $\kappa a_4(G) \geq 1$. Since $a_4(G) > 0$ by Theorem 2.47, the result follows. □

Corollary 4.22 *Suppose that* f *has a bounded Jacobian matrix, i.e.* $\left\|\frac{\partial f(x,t)}{\partial x}\right\| \leq M$ *for all* x, t. *Suppose also that* G *is a zero row sums matrix with nonpositive off-diagonal elements,* $D(t)$ *is symmetric and for some* $\epsilon > 0$, $D(t) \preceq -\epsilon I$ *for all* t. *Suppose further that* $\lim_{t\to\infty} \|u_i - u_j\| = 0$ *for all* i, j. *If the interaction graph contains a spanning directed tree, then the network in Eq. (4.8) synchronizes for sufficiently large* $\kappa > 0$.

Proof: If f has a bounded Jacobian matrix, we can choose $V = I$ and $f(x,t) + \psi D(t)x$ is V-uniformly decreasing for sufficiently large scalar ψ. The result then follows from Corollary 4.21. □

Corollaries 4.21-4.22 are intuitive results as the condition that the underlying graph contains a spanning directed tree implies that there exists a system that directly or indirectly couples into all other systems. So one can expect all other systems to synchronize to this system when the coupling is sufficiently large.

If the underlying graph does not contain a spanning directed tree, then the graph is not quasi-strongly connected by Theorem 2.1, i.e. there exists vertices i and j such that for all vertices k, either there is no path from k to i or no path from k to j. Consider the two groups of vertices with paths to i and j, denoted as $R(i)$ and $R(j)$ respectively. By definition, $R(i)$ and $R(j)$ are disjoint. Furthermore, there are no other vertices with paths to either $R(i)$ or $R(j)$. In other words, these are two groups of systems which are not influenced by other systems. Therefore in general these two groups of systems will not synchronize to each other, especially when the systems are chaotic and exhibit sensitive dependence on initial conditions. In this case the Frobenius normal form can be written as Eq. (2.2). Let V_i denote the set of systems corresponding to B_i and assume the conditions in Theorem 4.20 are satisfied. In this case the systems within V_j will synchronize with each other if $a_3(B_j) > 1$ for each $r \leq j \leq k$. Thus we have at least $k - r + 1$

separate clusters of synchronized systems. Similar arguments as above can be used to show that the systems belonging to $\cup_{i=1}^{r-1} V_i$ are synchronized to each other if for each $1 \leq j \leq r-1$, $\eta_j \geq 1$ and for each $r \leq j \leq k$, B_{ij} are constant row sums matrices with the row sum of B_{ij} equal to the row sum of $B_{i'j}$ for $1 \leq i < i' < r$.

In [Pecora and Carroll (1998b)] a Lyapunov exponents based approach is used to derive synchronization criteria. This method is based on numerical approximation of Lyapunov exponents and provides local results. The requirement that the underlying interaction graph contains a spanning directed tree also exists in the Lyapunov exponents approach to synchronization. In particular, it is a necessary condition since it implies that the zero eigenvalue is simple. In this approach, the synchronization criteria depend on the nonzero eigenvalue of G with the smallest real part. For chaotic systems, this eigenvalue needs to have a positive real part for the network to synchronize, a condition which is guaranteed by Theorem 2.9 or Corollary 2.12 when G is a Laplacian matrix.

4.3 Time-varying coupling topology

When the coupling topology $G(t)$ changes with time, the quantity $a_1(G)$ can be used to bound the coupling needed for synchronization.

Corollary 4.23 *The network of systems described by the state equations Eq. (4.1) synchronizes if*

(1) $\lim_{t\to\infty} \|u_i - u_j\| = 0$ *for all* i, j,
(2) $VD(t) = D^T(t)V \preceq 0$ *for all* t,
(3) There exists a symmetric positive definite matrix V *such that* $f(y,t) + a_1(G(t))D(t)y$ *is* V*-uniformly decreasing for all* t.

Proof: Apply Theorem 4.4 with $Y(t) = a_1(G(t))D(t)$ we get the condition

$$(U \otimes V)((G(t) - a_1(G(t))I) \otimes D(t)) \preceq 0$$

which is equivalent to $U(G(t) - a_1(G(t))I) \succeq 0$ since $VD(t) = D^T(t)V \preceq 0$. The proof of Theorem 4.8 show that $U = M$ as defined there satisfies this inequality. □

Note the difference between Corollary 4.6 and Corollary 4.23. In Corollary 4.6 the matrix U depends on G, whereas in Corollary 4.23 the matrix U is independent of $G(t)$. This is because in order to apply Theorem 4.4

the matrices U and V do not depend on time. In general, $a_1(G)$ is smaller than $\mu(G)$.

4.4 Coupling between delayed state variables

Consider a network of coupled systems with delay coupling by augmenting the state equation Eq. (4.1) with a term involving time-delayed state variables:

$$\dot{x} = \begin{pmatrix} f(x_1, t) \\ \vdots \\ f(x_n, t) \end{pmatrix} + (G(t) \otimes D(t))x + (G_\tau(t) \otimes D_\tau(t))x(t - \tau) + u(t) \quad (4.9)$$

We assume in this section that all matrices are real and denote the Moore-Penrose pseudoinverse of a matrix A as $A^\dagger$.

Lemma 4.24 *For matrices X and Y and a symmetric positive semidefinite matrix K of suitable dimensions,*

$$X^T KK^\dagger Y + Y^T KK^\dagger X \leq X^T KX + Y^T K^\dagger Y.$$

In particular, if x and y are vectors and K is symmetric positive definite, then $x^T y \leq \frac{1}{2} x^T Kx + \frac{1}{2} y^T K^{-1} y$.

Proof: Let the real Schur decomposition of K be $K = C^T \Gamma C$ where $C = C^{-T}$ is orthogonal and $\Gamma = \text{diag}(\lambda_1, \ldots, \lambda_n)$ is the diagonal matrix of eigenvalues. The Lemma then follows from

$$\begin{aligned} 0 &\leq (\sqrt{\Gamma} CX - \sqrt{\Gamma^\dagger} C^{-T} Y)^T (\sqrt{\Gamma} CX - \sqrt{\Gamma^\dagger} C^{-T} Y) \\ &\leq X^T C^T \Gamma CX + Y^T C^{-1} \Gamma^\dagger CY \\ &\quad - Y^T C^{-1} \sqrt{\Gamma^\dagger} \sqrt{\Gamma} CX - X^T C^T \sqrt{\Gamma} \sqrt{\Gamma^\dagger} C^{-T} Y \end{aligned}$$

as $C^{-1} \sqrt{\Gamma^\dagger} \sqrt{\Gamma} C = C^T \sqrt{\Gamma} \sqrt{\Gamma^\dagger} C^{-T} = KK^\dagger$ and $K^\dagger = C^{-1} \Gamma^\dagger C$. □

Lemma 4.24 can be further generalized as follows:

Lemma 4.25 *For matrices X and Y and a symmetric matrix K,*

$$X^T \sqrt{K^2} K^\dagger Y + Y^T \sqrt{K^2} K^\dagger X \leq X^T \sqrt{K^2} X + Y^T \sqrt{K^2}^\dagger Y.$$

Proof: As in Lemma 4.24 let the real Schur decomposition of K be $C^T \Gamma C$. Let $L = \text{diag}(\text{sign}(\lambda_1), \ldots, \text{sign}(\lambda_n))$ where $\text{sign}(x) = 1$ for $x \geq 0$ and -1 otherwise. Note that L is orthogonal and $\sqrt{K^2} = C^T \Gamma LC \geq 0$. By Lemma

4.24, $X^T\sqrt{K^2}\sqrt{K^2}^\dagger Y + Y^T\sqrt{K^2}\sqrt{K^2}^\dagger X \le X^T\sqrt{K^2}X + Y^T\sqrt{K^2}^\dagger Y$. Using the invertible transformation $Y \to C^{-1}LCY$ and the observation that $\sqrt{K^2}^\dagger C^{-1}LC = K^\dagger$ and $C^{-1}LC\sqrt{K^2}^\dagger C^{-1}LC = \sqrt{K^2}^\dagger$, we get the required inequality. □

The next result gives conditions under which the network in Eq. (4.9) synchronizes.

Theorem 4.26 *Let V be some symmetric positive definite matrix such that $f(y,t) + Y(t)y$ is V-uniformly decreasing. Let U be a matrix in $\mathcal{W}_s$, $(B_1(t), B_2(t))$ a factorization of $UG_\tau(t) \otimes VD_\tau(t) = B_1(t)B_2(t)$, and $K(t)$ a positive definite symmetric matrix for all t. The network synchronizes if $\|u_i - u_j\| \to 0$ as $t \to \infty$ and*

$$\begin{aligned} R \triangleq\, &(U \otimes V)(G(t) \otimes D(t) - I \otimes Y(t)) \\ &+\tfrac{1}{2}B_1(t)K(t)B_1^T(t) + \tfrac{1}{2}B_2^T(t)K^{-1}(t)B_2(t) \preceq 0 \end{aligned} \tag{4.10}$$

for all t.

Proof: Construct the Lyapunov functional $E(t) = \frac{1}{2}x(t)^T(U \otimes V)x(t) + \int_{t-\tau}^{t} x^T(s)U_\tau(s)x(s)ds$ where U_τ is a symmetric positive semidefinite matrix to be determined later. Note that $(U \otimes V) \succeq 0$. The derivative of E along trajectories of Eq. (4.9) is:

$$\begin{aligned} \dot{E} = &x^T(U \otimes V)(I \otimes f(x_i,t) + (I \otimes Y)x) \\ &+x^T(U \otimes V)(G \otimes D - I \otimes Y)x \\ &+x^T(U \otimes V)(G_\tau \otimes D_\tau)x(t-\tau) \\ &+x^T U_\tau x - x(t-\tau)^T U_\tau x(t-\tau) \end{aligned}$$

Using the same argument as in Theorem 4.4, we obtain

$$x^T(U \otimes V)(I \otimes f(x_i,t) + (I \otimes Y)x) \le -\mu x^T(U \otimes V)x \tag{4.11}$$

for some $\mu > 0$. Next, we use Lemma 4.24 to obtain:

$$\begin{aligned} &x^T(U \otimes V)(G_\tau \otimes D_\tau)x(t-\tau) = (x^T B_1)(B_2 x(t-\tau)) \\ &\le \tfrac{1}{2}x^T B_1 K B_1^T x^T + \tfrac{1}{2}x(t-\tau)^T B_2^T K^{-1} B_2 x(t-\tau) \end{aligned}$$

If we choose $U_\tau = \frac{1}{2}B_2^T K^{-1} B_2$ which is a symmetric positive semidefinite matrix for all t, then

$$\dot{E} \le -\mu x^T(U \otimes V)x + x^T R x$$

If $R \preceq 0$, then by Lyapunov's method [Hale (1977); Lakshmikantham and Liu (1993)] the trajectories approach the set $\{x : \dot{E} = 0\}$. If $\dot{E} = 0$, the above equation implies that $x^T(U \otimes V)x = 0$. Since $U \in \mathcal{W}_s$ and $V \succ 0$, by Lemma 2.18 this means that $x \in \mathcal{M}$. Therefore the set $\{x : \dot{E} = 0\}$ is a subset of the synchronization manifold $\mathcal{M}$ and thus the network synchronizes. □

The application of Theorem 4.26 has several degrees of freedom: the choice of (B_1, B_2), the choice of K and the choice of U. Next we study several of these choices that simplify the condition in Eq. (4.10).

4.4.1 *Choosing the factorization* $B_1B_2 = UG_\tau \otimes VD_\tau$

There are several ways to choose the factorization (B_1, B_2). Depending on the factorization, the matrix K can have a different dimension than $G \otimes D$ and $G_\tau \otimes D_\tau$. When the delay coupling term is absent ($G_\tau \otimes D_\tau = 0$), we can pick $B_1 = B_2 = 0$ and the synchronization criterion reverts back to the nondelay criterion in Theorem 4.4. The factorization should be chosen such that the synchronization manifold $\mathcal{M}$ is in the kernel of both B_1^T and B_2. Otherwise, as $\mathcal{M}$ is in the kernel of $(U \otimes V)$, this would mean that the matrix R in Eq. (4.10) is never negative semidefinite. Therefore if Eq. (4.10) is satisfied, then G_τ has constant row sums. This can be seen as follows. If G_τ does not have constant row sums, then $UG_\tau\mathbf{1} \neq 0$ and $\mathcal{M}$ is not in the kernel of $UG_\tau \otimes VD_\tau$ and thus also not in the kernel of B_2.

Let $J = \mathbf{1}\mathbf{1}^T$ be the matrix of all 1's and $M = I - \frac{1}{n}J \in \mathcal{W}_s$. Note that nM is the Laplacian matrix of the complete graph. The eigenvalues of M are 0 and 1. If X is a matrix with zero column sums, then $JX = 0$ and thus $MX = X$. In particular, $M^2 = M$, and $MU = UM = U$ for $U \in \mathcal{W}_s$. By choosing the factorizations $(B_1, B_2) = (U \otimes V, G_\tau \otimes D_\tau)$ and $(B_1, B_2) = (M \otimes I, UG_\tau \otimes VD_\tau)$ we get the following Corollary:

Corollary 4.27 *Let V be some symmetric positive definite matrix such that $f(y,t) + Y(t)y$ is V-uniformly decreasing. Let U be a matrix in $\mathcal{W}_s$ and $K(t)$ a positive definite symmetric matrix for all t. The network synchronizes if one of the following 2 conditions is satisfied for all t:*

$$\begin{aligned}&(U \otimes V)(G(t) \otimes D(t) - I \otimes Y(t))\\&+\tfrac{1}{2}(G_\tau(t) \otimes D_\tau(t))^T K^{-1}(t)(G_\tau(t) \otimes D_\tau(t))\\&+\tfrac{1}{2}(U \otimes V)K(t)(U \otimes V) \preceq 0\end{aligned} \tag{4.12}$$

$$\begin{aligned}&(U \otimes V)(G(t) \otimes D(t) - I \otimes Y(t))\\&+\tfrac{1}{2}(UG_\tau(t) \otimes VD_\tau(t))^T K^{-1}(t)(UG_\tau(t) \otimes VD_\tau(t))\\&+\tfrac{1}{2}(M \otimes I)K(t)(M \otimes I) \preceq 0\end{aligned} \tag{4.13}$$

The condition in Eq. (4.12) was obtained in [Amano *et al.* (2003)] for the case of constant coupling. If G_τ has zero row sums, we can choose the factorization $(B_1, B_2) = (UG_\tau \otimes VD_\tau, M \otimes I)$ to get:

Corollary 4.28 *Let V be some symmetric positive definite matrix such that $f(y,t) + Y(t)y$ is V-uniformly decreasing. Let U be a matrix in $\mathcal{W}_s$ and $K(t)$ a positive definite symmetric matrix for all t. If G_τ has zero row sums, then the network synchronizes if the following condition is satisfied for all t:*

$$\begin{aligned}&(U \otimes V)(G(t) \otimes D(t) - I \otimes Y(t)) + \tfrac{1}{2}(M \otimes I)K(t)(M \otimes I)\\&+\tfrac{1}{2}(UG_\tau(t) \otimes VD_\tau(t))K^{-1}(t)(UG_\tau(t) \otimes VD_\tau(t))^T \preceq 0\end{aligned}$$

4.4.2 Choosing the matrix $U \in \mathcal{W}_s$

By choosing $U = M$ and using the fact that $MX = X$ when X is a zero column sum matrix, the synchronization condition can be further simplified:

Corollary 4.29 *Let V be some symmetric positive definite matrix such that $f(y,t) + Y(t)y$ is V-uniformly decreasing. Let $K(t)$ be a positive definite symmetric matrix for all t. Suppose G_τ and G are zero column sums matrices. The network in Eq. (4.9) synchronizes if one of the following conditions is satisfied for all t:*

$$\begin{aligned}&G(t) \otimes VD(t) - M \otimes VY(t) + \tfrac{1}{2}(M \otimes V)K(t)(M \otimes V)\\&+\tfrac{1}{2}(G_\tau(t) \otimes D_\tau(t))^T K^{-1}(t)(G_\tau(t) \otimes D_\tau(t)) \preceq 0\end{aligned} \tag{4.14}$$

$$\begin{aligned}&G(t) \otimes VD(t) - M \otimes VY(t) + \tfrac{1}{2}(M \otimes I)K(t)(M \otimes I)\\&+\tfrac{1}{2}(G_\tau(t) \otimes VD_\tau(t))^T K^{-1}(t)(G_\tau(t) \otimes VD_\tau(t)) \preceq 0\end{aligned} \tag{4.15}$$

4.4.3 Choosing the matrix K

Corollary 4.30 *Let V be some symmetric positive definite matrix such that $f(y,t) + Y(t)y$ is V-uniformly decreasing. Suppose $G_\tau(t)$ and $VD_\tau(t)$ are symmetric for all t, and G and G_τ are zero column sums matrix. Suppose further that G_τ has a simple zero eigenvalue and D_τ is nonsingular for all t. The network in Eq. (4.9) synchronizes if the following condition is satisfied for all t:*

$$G(t) \otimes VD(t) - M \otimes VY(t) + \sqrt{(G_\tau \otimes VD_\tau)^2} \preceq 0 \tag{4.16}$$

In particular, if in addition $G_\tau(t) \otimes VD_\tau(t)$ is symmetric positive semidefinite for all t, then the network synchronizes if $G(t) \otimes VD(t) + G_\tau(t) \otimes VD_\tau(t) - M \otimes VY(t) \preceq 0$.

Proof: Since M commutes with G_τ, they can be simultaneously diagonalized, i.e. there exists a orthogonal matrix C such that $M = C^T\Gamma_M C$ and $G_\tau = C^T\Gamma_\tau C$ with Γ_M and Γ_τ diagonal matrices of the eigenvalues [Horn and Johnson (1985)]. Let $\Gamma_\tau = \mathrm{diag}(\lambda_1, \ldots, \lambda_n)$. By hypothesis, $\lambda_1 = 0$ and $\lambda_i > 0$ for $i > 1$. Let $H = C^T\mathrm{diag}(1, 0, \ldots, 0)C$ and $K = \sqrt{(G_\tau \otimes VD_\tau)^2} + H \otimes I$. It is easy to see that K is symmetric positive definite and $K^{-1} = \sqrt{(G_\tau \otimes VD_\tau)^2}^\dagger + H \otimes I$. Since $MH = G_\tau H = 0$ and $M\sqrt{(G_\tau \otimes VD_\tau)^2} = \sqrt{(G_\tau \otimes VD_\tau)^2}$, Eq. (4.15) reduces to Eq. (4.16). □

4.4.4 *The case $D = 0$*

Consider the case where all the coupling involves delayed state variables, i.e. $D = 0$. Assume that each individual system $\dot{x} = f(x,t)$ is globally asymptotically stable[2] such that the addition of positive feedback of the form $Y(t)x$ where $Y(t) \succeq 0$ does not change its stability. Then this stability is not destroyed if the delay coupling is small, i.e. each system x_i still converges towards the unique trajectory. More precisely, we have the following Corollary to Theorem 4.26.

Corollary 4.31 *Let V be some symmetric positive definite matrix such that $f(y,t) + Y(t)y$ is V-uniformly decreasing. Let U be a matrix in $\mathcal{W}_s$, $(B_1(t), B_2(t))$ a factorization of $UG_\tau(t) \otimes VD_\tau(t) = B_1(t)B_2(t)$, and $K(t)$ a positive definite symmetric matrix for all t. The network with state equations*

$$\dot{x}(t) = I \otimes f(x_i, t) + (G_\tau(t) \otimes D_\tau(t))x(t - \tau)$$

synchronizes if

$$U \otimes VY(t) \succeq \frac{1}{2}B_1(t)K(t)B_1^T(t) + \frac{1}{2}B_2^T(t)K^{-1}(t)B_2(t) \tag{4.17}$$

for all t.

---

[2] In the sense that for all trajectories x, and $\tilde{x}$ of the system, $x(t) - \tilde{x}(t) \to 0$ as $t \to \infty$.

4.5 Synchronization criteria based on algebraic connectivity

Theorem 4.32 *Let V be some symmetric positive definite matrix such that $f(y,t) + D(t)y$ is V-uniformly decreasing. The network in Eq. (4.9) synchronizes if the following conditions is satisfied for all t and some $\alpha(t) > 0$:*

$$F \triangleq (MG(t) - M) \otimes VD(t) + \tfrac{\alpha(t)}{2}(M \otimes I) + \tfrac{(MG_\tau(t) \otimes VD_\tau(t))^T (MG_\tau(t) \otimes VD_\tau(t))}{2\alpha(t)} \preceq 0 \tag{4.18}$$

If in addition $VD(t) \prec 0$ for all t and G is a zero row sums matrix, then the network synchronizes if

$$a_1(G(t)) \geq 1 + \|MG_\tau(t)\| \|VD_\tau(t)\| \|(VD(t))^{-1}\| \tag{4.19}$$

Proof: Eq. (4.18) follows from Corollary 4.27 by choosing $U = M$, $Y = D$ and $K = \alpha I$. Let us choose $\alpha(t) = \max(\|MG_\tau(t)\| \|VD_\tau(t)\|, \epsilon \cdot \nu(VD(t)))$ for some scalar $\epsilon > 0$ where $\nu(X) = \|X^{-1}\|^{-1}$ is the co-norm of the matrix X. For all z, $F(\mathbf{1} \otimes z) = 0$. For y a unit norm vector orthogonal to $\mathbf{1}$ and any unit norm vector z, define $w = y \otimes z$. Since $VD \prec 0$ and $My = y$,

$$\begin{aligned} w^T \left((MG(t) - M) \otimes VD(t)\right) w &\leq -(y^T G y - 1)\nu(VD(t)) \\ &\leq -(a_1(G) - 1)\nu(VD(t)) \end{aligned}$$

Furthermore, $\frac{\alpha}{2} w^T (M \otimes I) w \leq \frac{\alpha}{2}$ and

$$\tfrac{1}{2\alpha} w^T (MG_\tau \otimes VD_\tau)^T (MG_\tau \otimes VD_\tau) w \leq \tfrac{\|MG_\tau\|^2 \|VD_\tau\|^2}{2\alpha} \leq \tfrac{\alpha}{2}$$

This implies that F is negative semidefinite and that the network synchronizes if for any $\epsilon > 0$, $a_1(G)$ is larger than or equal to $1 + \max\left(\|MG_\tau(t)\| \|VD_\tau(t)\|, \epsilon \cdot \nu(VD(t))\right) \|(VD(t))^{-1}\|$. This last quantity is equal to $1 + \max(\|MG_\tau(t)\| \|VD_\tau(t)\| \|(VD(t))^{-1}\|, \epsilon)$. Combine this with the fact that $\mu > 0$ in Eq. (4.11) implies that the condition in Eq. (4.19) also synchronizes the network. □

When G is a zero row and column sums matrix, $MG = G$. Furthermore, $a_1(G) = \lambda_2^s(G)$. Therefore we have the following corollary to Theorem 4.32:

Corollary 4.33 *Let V be some symmetric positive definite matrix such that $f(y,t) + D(t)y$ is V-uniformly decreasing. Suppose G_τ and G are zero column sums matrices. The network in Eq. (4.9) synchronizes if the* *following condition is satisfied for all t and some $\alpha(t) > 0$:*

$$\begin{aligned}&(G(t) - M) \otimes VD(t) + \tfrac{\alpha(t)}{2}(M \otimes I)\\&+\tfrac{1}{2\alpha(t)}(G_\tau(t) \otimes VD_\tau(t))^T(G_\tau(t) \otimes VD_\tau(t)) \preceq 0\end{aligned} \quad (4.20)$$

If in addition $VD(t) \prec 0$ for all t and G is a zero row sums matrix, then the network synchronizes if

$$\lambda_2^s(G) \geq 1 + \|G_\tau(t)\|\|VD_\tau(t)\|\|(VD(t))^{-1}\| \quad (4.21)$$

When the nondelay coupling topology $G(t)$ does not change with time, Theorem 4.32 can be further improved:

Theorem 4.34 *Let V be some symmetric positive definite matrix such that $f(y,t) + D(t)y$ is V-uniformly decreasing. Let G be an irreducible zero row sums matrix and w be a positive vector such that $w^TG = 0$ and $\|w\|_\infty = 1$. Let W be a diagonal matrix with the vector w on the diagonal. Assume that $VD(t) \prec 0$ for all t. Then the network*

$$\dot{x}(t) = I \otimes f(x_i, t) + (G \otimes D(t))x(t) + (G_\tau(t) \otimes D_\tau(t))x(t - \tau) \quad (4.22)$$

synchronizes if

$$a_2(G) \geq 1 + \|UG_\tau(t)\|\|VD_\tau(t)\|\|(VD(t))^{-1}\|$$

where $U = W - \frac{ww^T}{\|w\|_1}$.

Proof: First note that $U \in \mathcal{W}_s$. By choosing $Y = D$ and $K = \alpha I$, Corollary 4.27 shows that the network synchronizes if $(UG - U) \otimes VD(t) + \frac{\alpha(t)}{2}(M \otimes I) + \frac{1}{2\alpha(t)}(UG_\tau(t) \otimes VD_\tau(t))^T(UG_\tau(t) \otimes VD_\tau(t)) \preceq 0$. Let us choose $\alpha(t) = \max(\|UG_\tau(t)\|\|VD_\tau(t)\|, \epsilon \cdot \nu(VD(t)))$ with $\epsilon > 0$. Let y be a unit norm vector orthogonal to $\mathbf{1}$. Note that

$$y^TU(G - I)y = y^TWGy - y^TUy \geq a_2(G) - y^TUy \geq a_2(G) - 1$$

since $\|U\| \leq 1$. Therefore

$$w^T\left((UG(t) - U) \otimes VD(t)\right)w \leq -(a_2(G) - 1)\nu(VD(t))$$

and the rest of the proof is similar to Theorem 4.32. □

Similar to Corollary 4.6 and Corollary 4.23, Theorem 4.32 and Theorem 4.34 relate synchronization of Eq. 4.9 to the algebraic connectivities $a_1(G)$ and $a_2(G)$. In particular, the network synchronizes if $a_1(G)$ is large enough for the time-varying case and $a_2(G)$ is large enough for the constant coupling topology case.

4.6 Further reading

More details about the results in this chapter can be found in [Wu and Chua (1995); Wu (2003a, 2005g,f, 2006a)]. See also [Wang and Chen (2002)] for a discussion on synchronization in networks of coupled systems. Another graph-theoretical approach to synchronization can be found in [Belykh *et al.* (2004b,a)].

Chapter 5

Synchronization in Networks of Coupled Discrete-time Systems

Consider the following state equation of a network of coupled discrete-time systems:

$$
\begin{aligned}
x(k+1) &= \begin{pmatrix} x_1(k+1) \\ \vdots \\ x_n(k+1) \end{pmatrix} \\
&= (I - G(k) \otimes D(k)) \begin{pmatrix} f(x_1(k), k) \\ \vdots \\ f(x_n(k), k) \end{pmatrix} + \begin{pmatrix} u_1(k) \\ \vdots \\ u_n(k) \end{pmatrix} \\
&= (I - G(k) \otimes D(k)) F(x(k), k) + u(k) \qquad (5.1)
\end{aligned}
$$

where

$$
u(k) = \begin{pmatrix} u_1(k) \\ \vdots \\ u_n(k) \end{pmatrix}.
$$

Synchronization in such a network of coupled discrete-time systems can be deduced from the eigenvalues of the coupling matrix G. The presentation here is adapted from [Wu (1998a, 2002, 2006b)].

Theorem 5.1 *Consider the network of coupled discrete-time systems with state equation Eq. (5.1) where $G(k)$ is a normal $n \times n$ matrix with zero row sums for each k. Let V be a symmetric positive definite matrix such that*

$$
(f(z,k) - f(\tilde{z},k))^T V (f(z,k) - f(\tilde{z},k)) \leq c^2 (z - \tilde{z})^T V (z - \tilde{z}) \qquad (5.2)
$$

for $c > 0$ and all $k, z, \tilde{z}$ and V has a decomposition[1] *of the form $V = C^T C$.*

[1] An example of such a decomposition is the Cholesky decomposition.

If

- $\|u_i(k) - u_j(k)\| \to 0$ *as* $k \to \infty$ *for all* i, j
- *For each* k, $\|I - \lambda CD(k)C^{-1}\|_2 < \frac{1}{c}$ *for every eigenvalue* λ *of* $G(k)$ *not corresponding to the eigenvector* $\mathbf{1}$.

then the coupled system synchronizes, i.e. $x_i(k) \to x_j(k)$ *for all* i, j *as* $k \to \infty$.

Proof: First consider the case $V = I$, i.e., $f(x, k)$ is Lipschitz continuous in x with Lipschitz constant c. Since $G(t)$ is normal, $G(t)$ has an orthonormal set of eigenvectors. Denote A as the subspace of vectors of the form $\mathbf{1} \otimes v$. Since $G\mathbf{1} = 0$ it follows that A is in the kernel of $G \otimes D$. Let b_i be the eigenvectors of G of unit norm corresponding to the eigenvalues λ_i in $L(G)$. Let us denote the subspace orthogonal to A by B. The vector $x(k)$ is decomposed as $x(k) = y(k) + z(k)$ where $y(k) \in A$ and $z(k) \in B$. By hypothesis $\|F(x(k), k) - F(y(k), k)\| \leq c\|z(k)\|$. We can decompose $F(x(k), k) - F(y(k), k)$ as $a(k) + b(k)$ where $a(k) \in A$ and $b(k) \in B$. Note that $F(y(k), k) - u(k)$ and $a(k)$ are in A and therefore $(G \otimes D)F(y(k), k) = (G \otimes D)u(k)$ and $(G \otimes D)a(k) = 0$.

$$\begin{aligned}
x(k+1) &= (I - G \otimes D)F(x(k), k) \\
&= (I - G \otimes D)(a(k) + b(k) + F(y(k), k)) \\
&= a(k) + F(y(k), k) + (I - G \otimes D)b(k) - (G \otimes D)u(k)
\end{aligned}$$

Since $a(k) \perp b(k)$, $\|b(k)\| \leq \|F(x(k), k) - F(y(k), k)\| \leq c\|z(k)\|$. Since $b(k) \in B$, it is of the form $b(k) = \sum_i \alpha_i b_i \otimes f_i(k)$.

$$\begin{aligned}
(I - G \otimes D)b(k) &= \sum_i \alpha_i b_i \otimes f_i(k) - \sum_i \alpha_i \lambda_i b_i \otimes D f_i(k) \\
&= \sum_i \alpha_i b_i \otimes (I - \lambda_i D) f_i(k)
\end{aligned}$$

Therefore

$$\begin{aligned}
\|(I - G \otimes D)b(k)\|^2 &= \sum_i |\alpha_i|^2 \|(I - \lambda_i D) f_i(k)\|^2 \\
&< \sum_i |\alpha_i|^2 \frac{\|f_i(k)\|^2}{c^2} = \frac{\|b(k)\|^2}{c^2} \leq \|z(k)\|^2
\end{aligned}$$

by hypothesis. Let $w(k)$ be the orthogonal projection of $(G \otimes D)u(k)$ onto B. Since $a(k) + F(y(k), k)$ is in A, $z(k+1) + w(k)$ is the orthogonal projection of $(I - G \otimes D)b(k)$ onto B and thus $\|z(k+1) + w(k)\| \leq \|(I - G \otimes D)b(k)\|$. By hypothesis, there exists $\beta < 1$ such that $\|(I - G \otimes D)b(k)\| \leq \frac{\beta}{c}\|b(k)\|$. Thus $\|z(k+1) + w(k)\| \leq \beta\|z(k)\|$. Note that $(G \otimes D)u(k) \to 0$ and thus also $w(k) \to 0$ as $k \to \infty$. This implies that $z(k) \to 0$ as $k \to \infty$. This means that $x(k)$ approaches the synchronization manifold $\mathcal{M}$.

For general V, let $C^T C$ be the decomposition of V. Applying the state transformation $y = (I \otimes C)x$, we get

$$\begin{aligned} y(k+1) &= (I \otimes C)(I - G \otimes D)F((I \otimes C^{-1})y(k), k) \\ &= (I - G \otimes CDC^{-1})(I \otimes C)F((I \otimes C^{-1})y(k), k) \\ &= (I - G \otimes CDC^{-1}) \begin{pmatrix} \tilde{f}(y_1(k), k) \\ \vdots \\ \tilde{f}(y_n(k), k) \end{pmatrix} \end{aligned}$$

where $\tilde{f}(y_i, k) = Cf(C^{-1}y_i, k)$. Then $\tilde{f}$ being Lipschitz continuous with Lipschitz constant c corresponds to Eq. (5.2) and the conclusion follows. □

When D is a normal matrix, Theorem 5.1 can be further simplified.

Theorem 5.2 *Consider the coupled network in Eq. (5.1) where G and D are normal matrices and G has zero row sums. Suppose f is Lipschitz continuous in x with Lipschitz constant c. If $|1 - \lambda\mu| < \frac{1}{c}$ for every eigenvalue λ of G in $L(G)$ and every eigenvalue μ of D then the coupled system (5.1) synchronizes.*

Proof: Since G and D are normal, so is $G \otimes D$. If $b(k) \in B$, by hypothesis $\|(I - G \otimes D)b(k)\| \leq \frac{\|b(k)\|}{c}$. The rest of the proof is similar to Theorem 5.1. □

A graphical interpretation of Theorem 5.2 is that the coupled system synchronizes if $L(G)$ lies in the interior of the intersection of circles of radii $\frac{1}{c|\mu_i|}$ centered at $\frac{1}{\mu_i}$ in the complex plane where μ_i are the eigenvalues of D. This is illustrated in Fig. 5.1. A dual interpretation is obtained by interchanging the roles of $L(G)$ and the eigenvalues of D.

When G and D are symmetric, their eigenvalues are real, and we have the following corollary:

Corollary 5.3 *Let c be the Lipschitz constant of f. The coupled system in Eq. (5.1) synchronizes if the following conditions are satisfied:*

- *D is symmetric and has only positive eigenvalues between μ_1 and μ_2 with $0 < \mu_1 \leq \mu_2$;*

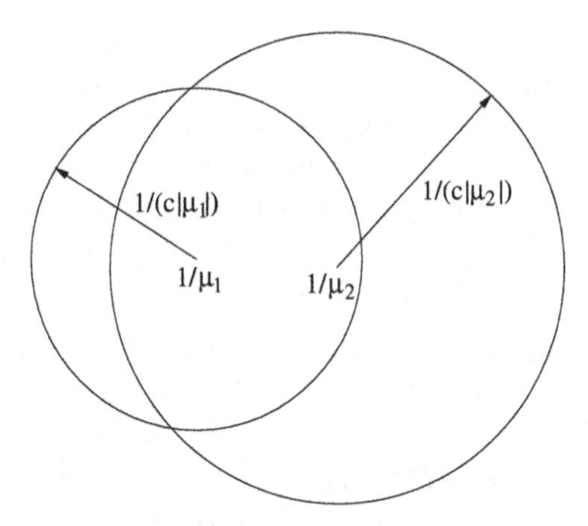

Fig. 5.1 Graphical interpretation of synchronization criterion for networks of coupled discrete time systems. Coupled network (5.1) synchronizes if all nonzero eigenvalues of G lie inside the intersection of the circles, where μ_i are the eigenvalues of D.

- *G is a symmetric matrix with zero row sums and a zero eigenvalue of multiplicity* 1*;*
- *the nonzero eigenvalues of G are in the interval* $\left(\frac{1-\frac{1}{c}}{\mu_1}, \frac{1+\frac{1}{c}}{\mu_2}\right)$.

Note the difference between Corollary 5.3 and the results in Chapter 4. When G is symmetric, in the continuous-time case the synchronization condition is a condition on the smallest (in magnitude) nonzero eigenvalue of G, while in the discrete-time case the synchronization condition is a condition on both the smallest and the largest nonzero eigenvalues of G. This is similar to the fact that the stability condition in the continuous time case (eigenvalues in the open left half plane) is mapped to the interior of the unit circle in the discrete time case via the mapping $z \to e^z$.

Consider the special case where the individual systems lie on the nodes of a weighted graph $\mathcal{G}$ and a uniform coupling exists between two systems if and only if there is an edge between the two corresponding nodes. The equation for the i-th system is given by:

$$x_i(k+1) = f(x_i(k), k) - \epsilon \sum_j a_{ij} D f(x_i(k), k) + \epsilon \sum_j a_{ij} D f(x_j(k), k) + u_i(k) \tag{5.3}$$

where $A = \{a_{ij}\}$ is the adjacency matrix of $\mathcal{G}$.

Corollary 5.4 *Let c be the Lipschitz constant of f and the graph $\mathcal{G}$ be* *connected. Let λ_2 and $\lambda_{\max}$ denote the smallest and largest nonzero eigenvalue of the Laplacian matrix L of $\mathcal{G}$ respectively. If*

- $\|u_i - u_j\| \to 0$ *as* $k \to \infty$ *for all* i, j
- D *is symmetric and has only positive eigenvalues between* μ_1 *and* μ_2 *(with* $0 < \mu_1 \leq \mu_2$*)*
- $\frac{\lambda_2}{\lambda_{\max}} > \frac{(c-1)\mu_2}{(c+1)\mu_1}$

then the system in Eq. (5.3) synchronizes with $\epsilon \in \left(\frac{1-\frac{1}{c}}{\mu_1\lambda_2}, \frac{1+\frac{1}{c}}{\mu_2\lambda_{\max}}\right)$.

Proof: In this case G is ϵL. The largest and smallest nonzero eigenvalues of G is $\epsilon\lambda_{\max}$ and $\epsilon\lambda_2$ respectively. Furthermore, since $\mathcal{G}$ is connected, the zero eigenvalue of G has multiplicity 1. The condition $\frac{\lambda_2}{\lambda_{\max}} > \frac{(c-1)\mu_2}{(c+1)\mu_1}$ implies that $\left(\frac{1-\frac{1}{c}}{\mu_1\lambda_2}, \frac{1+\frac{1}{c}}{\mu_2\lambda_{\max}}\right)$ is not empty. Then $\epsilon\lambda_2 > \frac{1-\frac{1}{c}}{\mu_1}$ and $\epsilon\lambda_{\max} < \frac{1+\frac{1}{c}}{\mu_2}$ as needed to apply Corollary 5.3. □

For the special case where D is a positive multiple of the identity matrix, $\mu_1 = \mu_2$ and Corollary 5.4 says that the network in Eq. (5.3) synchronizes for some ϵ if the ratio $r = \frac{\lambda_2}{\lambda_{\max}}$ is close enough to 1.

Example 5.5 Consider the case where $f(x) = ax(1-x)$ is the logistic map on the interval $[0,1]$ $(0 < a \leq 4)$. For $x \notin [0,1]$, we define $f(x) = f(x \bmod 1)$. Then $c = \sup_x |f'(x)| = |a|$. The case of n globally coupled logistic maps can be described by:

$$x_j(i+1) = (1-\epsilon)f(x_j(i)) + \frac{\epsilon}{n}\sum_{k=1}^{n} f(x_k(i))$$

for $j = 1, \ldots, n$ and $\epsilon > 0$. This system is of the form Eq. (5.1) where

$$G = \frac{\epsilon}{n}\begin{pmatrix} n-1 & -1 & -1 & \cdots & -1 \\ -1 & n-1 & -1 & \cdots & -1 \\ \vdots & & \ddots & & \vdots \\ -1 & -1 & \cdots & n-1 & -1 \\ -1 & -1 & \cdots & -1 & n-1 \end{pmatrix}$$

and $D = 1$. The eigenvalues of G are 0 and ϵ. Applying Theorem 5.2 we find that the system synchronizes if $|1-\epsilon| < \frac{1}{|a|}$.

This bound is more conservative than that obtained in [Kaneko (1989)] via numerical simulations. In Fig. 1 of [Kaneko (1989)], the curve which separates the synchronized state (the region which is denoted "COHERENT") from the other states lies below the curve $|1-\epsilon| = \frac{1}{|a|}$. However, both curves have the general shape with ϵ increasing as a is increased. In [Kaneko (1989)] it was shown that the local stability condition of the synchronized attractor is given by $|1-\epsilon| < \frac{1}{e^\lambda}$ where λ is the Lyapunov exponent of the logistic map.

Example 5.6 In [Chaté and Manneville (1992)] a lattice of n logistic maps is studied where each map is coupled to k logistic maps randomly chosen among the ensemble. In this case, it was found numerically that the system synchronizes for all large enough k. Assume in addition that the coupling is reciprocal, i.e. if there is a coupling between map i and map j, then there is an identical coupling between map j and map i. The corresponding G is then symmetric. There are exactly k nonzero elements in each row of $(I-G)$, each of them equal to $\frac{1}{k}$. Computer simulations indicate that for large n, G has one zero eigenvalue, and almost always the remaining eigenvalues lies in the range $\left[1-\frac{2}{\sqrt{k}}, 1+\frac{2}{\sqrt{k}}\right]$. Theorem 3.1 shows that for large n and even k this is true.

For coupling matrices G that are symmetric, Chapter 4 shows that the second smallest eigenvalue λ_2 of G provides an upper bound on the amount of coupling needed to synchronize the network. On the other hand, when the individual systems are discrete-time systems, Corollary 5.4 shows that under certain conditions the network of discrete time (5.1) synchronizes if the ratio $r = \frac{\lambda_2}{\lambda_{\max}}$ is large enough. This ratio r is also useful in characterizing synchronization thresholds obtained using the Lyapunov Exponent method for some matrices D in Eq. (4.1) [Pecora and Carroll (1998a,b); Barahona and Pecora (2002)]. The following result gives an upper and a lower bound on the ratio r.

Lemma 5.7 *For a connected graph of n vertices,*

$$\frac{\left(1-\cos\left(\frac{\pi}{n}\right)\right)\delta}{\Delta} \leq r \leq \frac{\delta}{\Delta}$$

where δ and Δ denotes the smallest and the largest vertex degree of the graph respectively.

Proof: Follows from Corollary 2.37 and Theorem 2.39. □

5.1 Synchronization of coupled scalar maps via contractivity of operators

Using the framework of contractive matrices we introduce in the next Chapter (Section 6.5) we can study synchronization in coupled map lattices of scalar maps, i.e. the case where the matrices $D(k)$ are constant scalars. Given a map $f_k : \mathbb{R} \rightarrow \mathbb{R}$, consider state variables $x_i \in \mathbb{R}$ that evolve according to f_k at time k: $x_i(k+1) = f_k(x(k))$. By coupling the output of these maps we obtain a coupled map lattice where each state evolves as:

$$x_i(k+1) = \sum_j a_{ij}(k) f_k(x_j(k))$$

This can be rewritten as

$$x(k+1) = A_k F_k(x(k)) \tag{5.4}$$

where $x(k) = (x_1(k), \ldots, x_n(k))^T \in \mathbb{R}^n$ is the vector of all the state variables and $F_k(x(k)) = (f_k(x_1(k)), \ldots, f_k(x_n(k)))^T$. We assume that A_k is a constant row sum matrix for all k. The map f_k depends on k, i.e. we allow the map in the lattice to be time varying. Furthermore, we do not require A_k to be a nonnegative matrix. If the row sum of A_k is 1, then this means that at synchronization, each state x_i in the lattice exhibits dynamics of the uncoupled map f_k, i.e. if $x(h) \in \mathcal{M} = \{\alpha \mathbf{1} : \alpha \in \mathbb{R}\}$, then for all $k \geq h$, $x(k) \in \mathcal{M}$ and $x_i(k+1) = f_k(x_i(k))$.

Theorem 5.8 *Let ρ_k be the Lipschitz constant of f_k. If $\lim_{k\rightarrow\infty} \prod_k c(A_k)\rho_k = 0$, where $c(A_k)$ is the set-contractivity with respect to $\mathcal{M}$ and a monotone norm, then the coupled map lattice in Eq. (5.4) synchronizes.*

Proof:

$$\|F_k(x(k)) - P(F_k(x(k))\| \leq \|F_k(x(k)) - F_k(P(x(k))\| \leq \rho_k \|x(k) - P(x(k)\|$$

where the last inequality follows from monotonicity of the norm. This implies that $c(F_k) \leq \rho_k$ and the result follows from Theorem 6.49. □

Thus we can synchronize the coupled map lattice if we can find matrices A_k and a norm such that the contractivities $c(A_k)$ are small enough. Let $r(A)$ denote the row sum of a constant row sum matrix A.

Corollary 5.9 *Let ρ_k be the Lipschitz constant of f_k. If* $\sup_k r(A_k) - \chi(A_k) - \frac{1}{\rho_k} < 0$, *then Eq. (5.4) synchronizes.*

Proof: Follows by applying Theorem 5.8 to set-contractivity with respect to $\|\cdot\|_\infty$. □

Chapter 6

Synchronization in Network of Systems with Linear Dynamics

In Chapters 4 and 5, the individual dynamical systems are *nonlinear* dynamical systems. When both the coupling and the individual systems are linear, the resulting network is a linear dynamical system which is simpler to analyze and allow us to say more about its synchronization properties.

6.1 Autonomous coupling

Consider affine state equations of the form:

$$\dot{x}(t) = \begin{pmatrix} \dot{x}_1(t) \\ \vdots \\ \dot{x}_n(t) \end{pmatrix} = (L \otimes D)x(t) + \mathbf{1} \otimes u(t) \tag{6.1}$$

for the continuous-time case and

$$x(k+1) = \begin{pmatrix} x_1(k+1) \\ \vdots \\ x_n(k+1) \end{pmatrix} (L \otimes D)x(k) + \mathbf{1} \otimes u(t) \tag{6.2}$$

for the discrete-time case. We assume that the row sums of A is 1 for Eq. (6.1) and 0 for Eq. (6.2). By using $y(t) = x(t) - \mathbf{1} \otimes \int_{-\infty}^{t} u(s)ds$, Eq. (6.1) can be reduced as[1]

$$\dot{y}(t) = (L \otimes D)y(t) \tag{6.3}$$

[1]Assuming that $\int_{-\infty}^{t} u(s)ds$ exists for all t.

Similarly, by using the recursive definition $p(k+1) = Dp(k) + u(k)$ and setting $y(k) = x(k) - \mathbf{1} \otimes p(k)$, Eq. (6.2) can be rewritten as

$$y(k+1) = (L \otimes D)y(k) \tag{6.4}$$

Note that the original systems Eqs. (6.1-6.2) synchronize if the reduced systems Eqs. (6.3-6.4) synchronize. In either case, standard linear algebra tells us that synchronization can determined by the location of the eigenvalues of L and D. In particular, the network in Eq. (6.1) synchronizes if all eigenvalues of D has negative real parts, the zero eigenvalue of L is simple and all other eigenvalues of L have positive real parts. Similarly, the network in Eq. (6.2) synchronizes if all eigenvalues of D has norm strictly less than 1, 1 is a simple eigenvalue of L and all other eigenvalues of L have norm strictly less than one.

Theorem 6.1 *If L is a zero row sum matrix with nonpositive off-diagonal elements and all eigenvalues of D has negative real parts, then the network in Eq. (6.1) synchronizes if the interaction graph of L contains a spanning directed tree.*

Proof: Follows from Corollary 2.25. □

Theorem 6.2 *If L is a stochastic matrix with nonzero diagonal elements and all eigenvalues of D has norm strictly less than 1, then the network in Eq. (6.2) synchronizes if the interaction graph of L contains a spanning directed tree.*

Proof: Follows from the fact that a consequence of Corollaries 2.24 and 2.27 is that the eigenvalue 1 of a stochastic matrix is simple if and only if the interaction graph of the matrix contains a spanning directed tree. □

6.2 Nonautonomous coupling: continuous-time case

The state equations are given by:

$$\dot{x} = (L(t) \otimes D(t))x(t) + \mathbf{1} \otimes u(t) \tag{6.5}$$

Theorem 6.3 *Suppose $L(t)$ is a zero row sum matrix at each time t, $D(t)$ is symmetric with eigenvalues less than $\epsilon < 0$ for all t. Then Eq. (6.5) synchronizes if there exists $\delta > 0$ such that $a_1(L(t)) \geq \delta$ for all t.*

Proof: This is a special case of Corollary 4.23 where we choose $V = I$ and $f = 0$. □

6.2.1 *Slowly varying coupling*

In contrast to the autonomous case, the coupled system with state equations Eq. (6.5) does not necessarily synchronize even if the interaction graph of $L(t)$ contains a spanning directed tree for each t, unless $L(t)$ changes slowly or the changes in $L(t)$ are small with respect to t, as will be shown next.

Consider the following result on exponential stability of nonautonomous linear systems [Ilchmann *et al.* (1987)]:

Theorem 6.4 *Let $A(t)$ be piecewise continuous as a function of time t and there exists $\alpha > 0$ such that all eigenvalues of $A(t)$ has real part less than $-\alpha$ for all $t \geq 0$. Suppose there exists $m > 0$ such that $\|A(t)\| \leq m$ for all $t \geq 0$. Then $\dot{x} = A(t)x(t)$ is exponentially stable if if one of the following conditions is satisfied for all $t \geq 0$:*

(1) $\alpha > 4m$;
(2) $A(\cdot)$ *is piecewise differentiable and*

$$\|\dot{A}(t)\| \leq \delta < \frac{\alpha^{4n-2}}{(2n-1)m^{4n-4}}$$

(3) For some $k \geq 0$, $0 < \eta < 1$, $\alpha > 4m\eta - \frac{n-1}{k}\log\eta$ *and*

$$\sup_{0 \leq \tau \leq k} \|A(t+\tau) - A(t)\| \leq \delta < \eta^{n-1}\left(\alpha - 4m\eta + \frac{n-1}{k}\log\eta\right)$$

(4) $\alpha > n-1$ *and for some* $0 < \eta < 1$

$$\sup_{h>0}\left\|\frac{A(t+h) - A(t)}{h}\right\| \leq \delta < 2\eta^{n-1}(\alpha - 2m\eta + (n-1)\log\eta)$$

Theorem 6.5 *Let $L(t)$ and $D(t)$ be piecewise continuous functions of time t and there exists $\alpha, \beta > 0$ such that all nonzero eigenvalues of $L(t)$ has real part larger than α and all eigenvalues of $D(t)$ has real part less than $-\beta$ for all $t \geq 0$. The Eq. (6.1) synchronizes if $L(t)$ and $D(t)$ changes slowly or the changes in $L(t)$ and $D(t)$ are small.*

Proof: Let C and R be as defined in Lemma 4.16 and $y_i = x_i - x_{i+1}$ for $i = 1, \ldots n-1$. Then $y = (C \otimes I)x$ and thus $\dot{y} = (CL(t) \otimes D(t))x$. Lemma 4.7 shows that $CL(t) = A(t)C$ where $A(t) = CL(t)R$ and $\gamma(L(t)) = \min_i\{\mathrm{Re}(\lambda_i) : \lambda_i \in \mathrm{Spectrum}(A(t))\}$. Thus $\dot{y} = (A(t) \otimes D(t))y$, and agreement is achieved if $y \to 0$. The result is then proved by applying Theorem 6.4. □

6.3 Nonautonomous coupling: discrete-time case

In the discrete time case, the state equations are written as:

$$x(k+1) = (M(k) \otimes D(k))x(k) + \mathbf{1} \otimes u(t) \tag{6.6}$$

We make the additional assumption that $\|D(k)\| \leq 1$ for all k and $M(k)$ are stochastic matrices, i.e. nonnegative matrices such that each row sums to 1. First we show that the maximum distance between the x_i's is nonincreasing:

Theorem 6.6 *Let $\kappa(k) = \max_{i,j} \|x_i(k) - x_j(k)\|$. Then $\kappa(k+1) \leq \kappa(k)$.*

Proof: Let $S(k) = \{x_1(k), \ldots, x_n(k)\}$ and $T(k) = \{D(k)x_1(k) + v(k), \ldots, D(k)x_n(k) + v(k)\}$. Note that $\kappa(k)$ is the diameter of the convex hull of $S(k)$. Since $x_i(k+1)$ is a convex combination of elements of $T(k)$, it is in the convex hull of $T(k)$. Thus means that the convex hull of $S(k+1)$ is a subset of the convex hull of $T(k)$. Since $\|D(k)\| \leq 1$, the diameter of the convex hull of $T(k)$ is less than or equal to the diameter of the convex hull of $S(k)$ and thus $\kappa(k+1) \leq \kappa(k)$. □

6.3.1 *A discrete-time consensus problem*

A group of agents would like to agree on a probability distribution of a (discrete) random variable ϕ. Each agent has a probability distribution in mind and the question is how to integrate information about other agents' probability distributions to reach consensus [Winkler (1968)]. In [DeGroot (1974)] this problem was studied and a procedure was proposed where each agent updates its guess by computing a weighted average of the distributions of every agent. Let $p_i(k)$ be the probability row vector of the i-th agent at time k and let $P(k)$ be the matrix consisting of the probability row vectors of all the agents at time k. The updating procedure can then be expressed as $P(k+1) = M(k)P(k)$ for some stochastic matrix $M(k)$. Iterating this, we see that we reach consensus if $M(n)M(n-1)\ldots M(1)$ converges to a rank one matrix $\mathbf{1}c^T$ as $n \to \infty$. In fact, the equation $P(k+1) = M(k)P(k)$ can be viewed as a special case of Eq. (6.6) by setting $D(k) = I$, $u(t) = 0$ and rearranging the rows of $P(k)$ into a single column vector $x(k)$.

6.4 Ergodicity of inhomogeneous Markov chains

It is clear that the system in Eq. (6.6) synchronizes if $\lim_{n\to\infty} M(n)M(n-1)\cdots M(1)$ converges to a rank one matrix of the form $\mathbf{1}c^T$. This is related to weak ergodicity of inhomogeneous Markov chains. Let $M(k)$ be the transition matrix of a Markov chain at time k.

Definition 6.7 [Hajnal (1958)] An inhomogeneous Markov chain with transition matrices $M(k)$ is *strongly ergodic* if $M(1)M(2)\cdots M(n)$ converges to a rank one matrix of the form $\mathbf{1}c^T$ as $n \to \infty$. The chain is *weakly ergodic* if the rows of $A = M(1)M(2)\cdots M(n)$ approach each other as $n \to \infty$, i.e., $\max_{i,j,k} |A_{ik} - A_{jk}| \to 0$ as $n \to \infty$.

Strong ergodicity implies weak ergodicity, but not vice versa in general. However, when the matrices are arranged in reverse order as is the case in analyzing Eq. (6.6), then strong ergodicity is equivalent to weak ergodicity in products of stochastic matrices.

Lemma 6.8 *[Chatterjee and Seneta (1977)] Let $M(i)$ be a set of stochastic matrices. $M(n)M(n-1)\cdots M(1)$ converges to a rank one matrix of the form $\mathbf{1}c^T$ as $n \to \infty$ if and only if the rows of $M(n)M(n-1)\cdots M(1)$ approach each other as $n \to \infty$.*

Proof: One direction is clear. Let $A(n) = M(n)M(n-1)\cdots M(1)$. Suppose the rows of $A(n)$ approach each other as $n \to \infty$, i.e. the convex hull of the rows of $A(n)$ is shrinking to a single point that is not necessarily fixed. Since each row of $A(n+1)$ is a convex combination of the rows of $A(n)$, the convex hull of the rows of $A(n+1)$ is a subset of the convex hull of $A(n)$ and thus converges to the intersection of the convex hulls which is a fixed single point, i.e. $A(n)$ converges to $\mathbf{1}c^T$. □

Definition 6.9 The ergodicity coefficient of a real matrix A is defined as $\chi(A) = \min_{j,k} \sum_i \min(A_{ji}, A_{ki})$.

Definition 6.10 A stochastic matrix is called *Markov* if it has a positive column. A stochastic matrix is called *scrambling* if for each pair of indices j, k, there exists i such that A_{ji} and A_{ki} are both nonzero.

It is clear that a Markov matrix is scrambling. Furthermore, for stochastic matrices $0 \le \chi(A) \le 1$ with $\chi(A) > 0$ if and only if A is scrambling.

Definition 6.11 For a real matrix A, define $\delta(A) \ge 0$ as

$$\delta(A) = \max_{i,j} \sum_k \max(0, A_{ik} - A_{jk}) \ge \max_{i,j,k}(A_{ik} - A_{jk})$$

Lemma 6.12 *If A is a matrix with constant row sums, then* $\delta(A) = \frac{1}{2}\max_{i,j}\sum_k |A_{ik} - A_{jk}|$.

Proof: For a fixed i and j, let K be the set of indices k such that $A_{ik} > A_{jk}$. Then $\frac{1}{2}\sum_k |A_{ik} - A_{jk}| = \frac{1}{2}\sum_{k\in K} A_{ik} - A_{jk} + \frac{1}{2}\sum_{k\notin K} A_{jk} - A_{ik}$. Since $\sum_k A_{ik} = \sum_k A_{jk}$, $\sum_{k\notin K} A_{jk} - A_{ik} = \sum_{k\in K} A_{ik} - A_{jk}$ and thus $\frac{1}{2}\sum_k |A_{ik} - A_{jk}| = \sum_{k\notin K} A_{ik} - A_{jk} = \sum_k \max(0, A_{ik} - A_{jk})$. □

The following result relates $\delta(A)$ to the ergodicity coefficient $\chi(A)$.

Theorem 6.13 *If A is a matrix where each row sum is r, then $\delta(A) = r - \chi(A)$. If A is a matrix where each row sum is less than or equal to r, then $\delta(A) \leq r - \chi(A)$.*

Proof: We follow the proof in [Paz (1970)]. For the first statement,

$$\begin{aligned}\delta(A) &= \max_{i,j}\sum_k \max(0, A_{ik} - A_{jk}) = \max_{i,j}\sum_k (A_{ik} - \min(A_{ik}, A_{jk})) \\ &= \max_{i,j}\sum_k (A_{ik} - \min(A_{ik}, A_{jk})) = r - \min_{i,j}\sum_k \min(A_{ik}, A_{jk}) \\ &= r - \chi(A) \end{aligned} \tag{6.7}$$

For the second statement

$$\begin{aligned}\delta(A) &= \max_{i,j}\sum_k (A_{ik} - \min(A_{ik}, A_{jk})) \\ &\leq \max_{i}\sum_k A_{ik} - \min_{i,j}\sum_k \min(A_{ik}, A_{jk}) \leq r - \chi(A) \end{aligned} \tag{6.8}$$

□

Theorem 6.14 *If A is a real matrix with constant row sums and $x \in \mathbb{R}^n$, then $\max_i y_i - \min_i y_i \leq \delta(A)(\max_i x_i - \min_i x_i)$ where $y = Ax$.*

Proof: The proof is similar to the argument in [Paz and Reichaw (1967)]. Let $x_{\max} = \max_i x_i$, $x_{\min} = \min_i x_i$, $y_{\max} = \max_i y_i$, $y_{\min} = \min_i y_i$.

$$\begin{aligned} y_{\max} - y_{\min} &= \max_{i,j}\textstyle\sum_k (A_{ik} - A_{jk})\,x_k \\ &\leq \max_{i,j}(\textstyle\sum_k \max(0, A_{ik} - A_{jk})\,x_{\max} \\ &\quad + \textstyle\sum_k \min(0, A_{ik} - A_{jk})\,x_{\min}) \end{aligned} \tag{6.9}$$

Since A has constant row sums, $\sum_k A_{ik} - A_{jk} = 0$, i.e.

$$\sum_k \max(0, A_{ik} - A_{jk}) + \sum_k \min((0, A_{ik} - A_{jk}) = 0$$

This means that

$$\begin{aligned} y_{\max} - y_{\min} &\leq \max_{i,j} \left(\sum_k \max\left(0, A_{ik} - A_{jk}\right)\right) (x_{\max} - x_{\min}) \\ &\leq \delta(A)\,(x_{\max} - x_{\min}) \end{aligned} \tag{6.10}$$

□

Classical conditions for an inhomogeneous Markov chain to be weakly ergodic are that the transition matrices are scrambling or Markov [Hajnal (1958); Seneta (1973)]. This follows from the following Lemma originally due to Hajnal and generalized by Dobrusin and by Paz and Reichaw:

Lemma 6.15 *[Hajnal (1958); Paz and Reichaw (1967)] If A_i are stochastic matrices, then $\delta(A_1 A_2 \ldots A_n) \leq \Pi_i(1 - \chi(A_i))$.*

Proof: We follow the presentation in [Paz and Reichaw (1967)] and show that $\delta(AB) \leq \delta(A)\delta(B)$ and the conclusion then follows from Theorem 6.13.

$$\begin{aligned} \delta(AB) &= \max_{i,j} \sum_k \max\left(0, \sum_l (A_{il} - A_{jl}) B_{lk}\right) \\ &= \max_{i,j} \sum_{k \in K} \sum_l (A_{il} - A_{jl}) B_{lk} \end{aligned} \tag{6.11}$$

where K is the set of indices k such that $\sum_l (A_{il} - A_{jl}) B_{lk} \geq 0$. Thus

$$\delta(AB) \leq \max_{i,j} \sum_{l \in L} (A_{il} - A_{jl}) \max_l \sum_{k \in K} B_{lk} + \sum_{l \notin L} (A_{il} - A_{jl}) \min_l \sum_{k \in K} B_{lk}$$

where L are the indices l such that $A_{il} - A_{jl} \geq 0$. Since A is a stochastic matrix, $\sum_{l \in L}(A_{il} - A_{jl}) = -\sum_{l \notin L}(A_{il} - A_{jl})$ and thus

$$\begin{aligned} \delta(AB) &\leq \max_{i,j} \sum_{l \in L} (A_{il} - A_{jl}) \left(\max_l \sum_{k \in K} B_{lk} - \min_l \sum_{k \in K} B_{lk} \right) \\ &\leq \delta(A) \left(\max_l \sum_{k \in K} B_{lk} - \min_l \sum_{k \in K} B_{lk} \right) \end{aligned} \tag{6.12}$$

since $\max_l \sum_{k \in K} B_{lk} - \min_l \sum_{k \in K} B_{lk}) = \max_{l_1, l_2} \sum_{k \in K} (B_{l_1 k} - B_{l_2 k}) \leq \max_{l_1, l_2} \sum_k \max(0, B_{l_1 k} - B_{l_2 k}) = \delta(B)$, we have $\delta(AB) \leq \delta(A)\delta(B)$. □

Lemma 6.16 *Let $0 \leq p_i \leq 1$. If $\sum_i p_i$ diverges then $\prod_i (1 - p_i) \to 0$. If in addition $p_i < 1$, then $\sum_i p_i$ diverges if and only if $\prod_i (1 - p_i) \to 0$.*

Proof: First note that $\prod_i(1-p_i)$ converges. Suppose $0 \leq p_i \leq 1$ and $\sum_i p_i$ diverges. Since $-\log(1-p) \geq p$, $-\log\prod_i(1-p_i)$ also diverges, i.e. $\prod_i(1-p_i) \to 0$. Suppose $0 \leq p_i < 1$ and $\sum_i p_i$ converges. Then $J = \{i : p_i > \frac{1}{2}\}$ is a finite set. This means that $\prod_{i\in J}(1-p_i)$ is nonzero. If $i \notin J$, then $-\log(1-p_i) < 2p_i$ and thus $-\log\prod_{i\notin J}(1-p_i)$ converges. This implies that $\prod_{i\notin J}(1-p_i)$ does not converge to 0. □

Lemma 6.16 is a well-known result according to [Hajnal (1958)], although the additional condition $p_i < 1$ needs to be added to [Hajnal (1958)] as it is not true otherwise. We were not successful in finding the original source and the proof here is due to Don Coppersmith.

Theorem 6.17 *The Markov chain with transition matrices $M(i)$ is weakly ergodic if $\sum_i \chi(M(i)) = \infty$.*

Proof: Follows from Lemmas 6.15 and 6.16. □

As we will show in Chapter 7, a consequence of Lemma 6.15 is that the magnitude of χ is important to determining how fast a network of discrete-time systems synchronize. This is analogous to the magnitude of the algebraic connectivity determining how fast a network of continuous-time systems synchronize. In fact, there is an analogous statement to Theorem 2.38: a union of imploding star graphs will maximize the ergodicity coefficient χ. For a graph $\mathcal{G}$, we construct a corresponding stochastic matrix $A^{\mathcal{G}}$ as follows. For each vertex v with outdegree $d_o(v)$, $A^{\mathcal{G}}_{vv} = A^{\mathcal{G}}_{vw} = \frac{1}{d_o(v)+1}$ for each vertex w with an edge from v to w. We use the notation $\chi(\mathcal{G})$ to denote $\chi(A^{\mathcal{G}})$.

Theorem 6.18 *[Cao and Wu (2007)] Consider the set Q of simple graphs with n vertices and $k(n-1)$ edges where $k < n$. Let $\mathcal{G}_k \in Q$ be a union of imploding star graphs (Fig. 2.7). Then $\chi(\mathcal{G}_k) = \frac{k}{k+1}$ is maximal among the elements of the set Q.*

Proof: It is easy to show that $\chi(\mathcal{G}_k) = \frac{k}{k+1}$. Suppose there exists a graph $\mathcal{H} \in Q$ such that $\chi(\mathcal{H}) > \frac{k}{k+1}$. Let u and v be indices such that $\chi(A^{\mathcal{H}}) = \sum_i \min\left(A^{\mathcal{H}}_{ui}, A^{\mathcal{H}}_{vi}\right)$. Without loss of generality, we assume $d = d_i(u) \geq d_i(v)$, where $d_i(v)$ denotes the indegree of the vertex v. Then each nonzero element in row u is $\frac{1}{d+1}$. Let $q(i,j)$, $i \neq j$, denote the number of columns of $A^{\mathcal{H}}$ each of which has positive elements intersecting with both row i and row j.

From the assumption $\chi(\mathcal{H}) > \frac{k}{k+1}$ and the selection of u and v, we know that for any w

$$\sum_i \min\left(A^{\mathcal{H}}_{ui}, A^{\mathcal{H}}_{wi}\right) \geq \sum_i \min\left(A^{\mathcal{H}}_{ui}, A^{\mathcal{H}}_{vi}\right) > \frac{k}{k+1} \tag{6.13}$$

On the other hand, we have

$$\sum_i \min\left(A^{\mathcal{H}}_{ui}, A^{\mathcal{H}}_{wi}\right) \leq q(u,w)\frac{1}{d+1} \leq \frac{d_i(w)+1}{d+1} \tag{6.14}$$

If we choose w to be the vertex of the minimum in-degree in $\mathcal{H}$, then

$$d_i(w) \leq \left\lfloor \frac{k(n-1)}{n} \right\rfloor = k-1 \tag{6.15}$$

Combining inequalities (6.13), (6.14) and (6.15) together, we have $\frac{k}{d+1} > \frac{k}{k+1}$, which implies $d < k$. Then there are at most k non-zero elements in row u. Consequently, for each l with $l \neq u$ and $l \neq v$, we have

$$\frac{k}{d_i(l)+1} \geq \sum_i \min\left(A^{\mathcal{H}}_{ui}, A^{\mathcal{H}}_{li}\right) > \frac{k}{k+1} \tag{6.16}$$

where the second inequality sign holds because of (6.13).

The inequality (6.16) implies that

$$d_i(l) < k, \qquad l \neq u \text{ and } l \neq v$$

Combined with the fact that $d_i(v) \leq d_i(u) = d < k$, we have proved that all the vertices of the graph $\mathcal{H}$ have indegree strictly less than k. So there are at most $(k-1)n$ edges in $\mathcal{H}$ which contradicts the fact that $\mathcal{H}$ has $k(n-1)$ edges. Hence, a graph $\mathcal{H}$ satisfying $\chi(\mathcal{H}) > \frac{k}{k+1}$ does not exist. □

In [Seneta (1973)] a more general definition of *coefficient of ergodicity* is given.

Definition 6.19 A coefficient of ergodicity is a continuous function χ on the set of n by n stochastic matrices such that $0 \leq \chi(A) \leq 1$. A coefficient of ergodicity χ is *proper* if

$$\chi(A) = 1 \Leftrightarrow A = \mathbf{1}v^T \quad \text{for some probability vector } v.$$

Ref. [Seneta (1973)] gives the following necessary and sufficient conditions for weak ergodicity generalizing the arguments by Hajnal.

Theorem 6.20 *Suppose χ_1 and χ_2 are coefficients of ergodicity such that χ_1 is proper and the following equation is satisfied for some constant C and all k,*

$$1 - \chi_1(S_1 S_2 \cdots S_k) \leq C \prod_{i=1}^{k}(1 - \chi_2(S_i)) \tag{6.17}$$

where S_i are stochastic matrices. Then a sequence of stochastic matrices A_i is weakly ergodic if there exists a strictly increasing subsequence $\{i_j\}$ such that

$$\sum_{j=1}^{\infty} \chi_2(A_{i_j+1} \cdots A_{i_{j+1}}) = \infty \tag{6.18}$$

Conversely, if A_i is a weakly ergodic sequence, and χ_1, χ_2 are both proper coefficients of ergodicity satisfying Eq. (6.17), then Eq. (6.18) is satisfied for some strictly increasing sequence $\{i_j\}$.

In [Seneta (1979)] a class of coefficients of ergodicity was defined with respect to a norm $\|\cdot\|$:

$$\chi_{\|\cdot\|}(A) = 1 - \sup_{x \perp \mathbf{1}, \|x\|=1} \|x^T A\|$$

$\chi_{\|\cdot\|}(A)$ is a proper coefficient of ergodicity and satisfies Eq. (6.17) with $C = 1$. Explicit forms are known for the norms $\|\cdot\|_1$, $\|\cdot\|_2$ and $\|\cdot\|_\infty$. $\chi_{\|\cdot\|_1}$ coincides with Definition 6.9, $\chi_{\|\cdot\|_2}(A) = 1 - \sqrt{a_1(AA^T)}$ and $\chi_{\|\cdot\|_\infty}$ is equal to [Tan (1982)]:

$$\chi_{\|\cdot\|_\infty}(A) = 1 - \max_{\eta} \min_{k} \sum_{j=1}^{n} |\eta_j - \eta_k|$$

where the maximum is taken over all columns η of A.

Definition 6.21 A nonnegative matrix A is primitive if and only if A^m is positive for some $m > 0$.

A nonnegative matrix A is primitive if and only if it is irreducible and has only one eigenvalue of maximum modulus [Minc (1988)].

Definition 6.22 A stochastic matrix A is stochastic, indecomposable and aperiodic (SIA) if $\lim_{n\rightarrow} A^n = \mathbf{1}c^T$ for some vector c.

A note of caution here about notations as they are inconsistent between literature in Markov chain theory and literature in other areas of mathematics. For instance, in Markov chain analysis, primitive matrices are also called regular matrices. On the other hand, some authors have referred to nonsingular matrices as regular matrices. Furthermore, in Markov chain theory, a square matrix is decomposable if it can be written (after simultaneous row and column permutations) in block diagonal form. This is not the same concept as reducibility (as defined in Section 2.2). However, some authors have defined decomposability of a matrix as being the same as reducibility. For consistency, we adhere to the definitions commonly used in the analysis of Markov chains.

For nonnegative matrices, the property of being primitive or being SIA depends only on the location of the nonzero entries of the matrix and not on the actual values of these entries. A stochastic primitive matrix is SIA, but products of primitive matrices are not necessarily primitive nor SIA as the following example adapted from [Hajnal (1958)] illustrates:

$$\begin{pmatrix} 0 & 1 & 0 \\ 0 & 0 & 1 \\ \frac{1}{3} & \frac{1}{3} & \frac{1}{3} \end{pmatrix} \begin{pmatrix} \frac{1}{3} & \frac{1}{3} & \frac{1}{3} \\ 1 & 0 & 0 \\ 0 & 1 & 0 \end{pmatrix} = \begin{pmatrix} 1 & 0 & 0 \\ 0 & 1 & 0 \\ \frac{4}{9} & \frac{4}{9} & \frac{1}{9} \end{pmatrix}$$

In [Wolfowitz (1963)] a more general sufficient condition for weak ergodicity is given:

Theorem 6.23 *If the transition matrices are from a finite set and finite products of transition matrices are SIA, then the Markov chain is weakly ergodic.*

In particular, [Wolfowitz (1963)] shows that:

Lemma 6.24 *If S is a set of stochastic matrices whose finite products are SIA, then there exists an integer n_W such that a finite product of matrices in S of length at least n_W is scrambling.*

Proof: By Theorem 6.13 and Lemma 6.15,

$$(1 - \chi(AB)) \leq (1 - \chi(A))\,(1 - \chi(B))$$

for stochastic matrices A and B. This implies that if B is scrambling, then so is AB. Let A_i be matrices in S. The locations of the nonzero entries of a stochastic matrix of fixed order can only take on a finite number of patterns. Thus a sequence of stochastic matrices products $A_1 A_2 \ldots$ will have two matrix products with the same pattern for the nonzero entries, say $A_1 A_2 \ldots A_k$ and $A_1 A_2 \ldots A_k \ldots A_m$. Let $B = A_{k+1} \ldots A_m$. By induction, $A_1 A_2 \ldots A_k B^i$ has the same pattern as $A_1 A_2 \ldots A_k$ for all i. Since B is an SIA matrix, for sufficiently large i, B^i is a scrambling (in fact, Markov) matrix. This means that $A_1 A_2 \ldots A_k B^i$ and thus $A_1 A_2 \ldots A_k$ is scrambling. Since k is bounded from above and S is finite, $\chi(A_1 A_2 \ldots A_k)$ will be bounded away from 0 and thus $\delta(A_1 A_2 \ldots A_i) \to 0$ as $i \to \infty$ by Lemma 6.15. □

Theorem 6.23 then follows from Lemma 6.24 and Lemma 6.15. In [Anthonisse and Tijms (1977)] it was shown that the conditions in Theorem 6.23 implies synchronization of Eq. (6.6):

Theorem 6.25 *If $M(k)$ are from a finite set and finite products of $M(k)$ are SIA, then $\lim_{n\to\infty} M(n)M(n-1)\cdots M(1)$ converges to a rank one matrix of the form $\mathbf{1}c^T$.*

Proof: Follows from Lemma 6.8 and Theorem 6.23. □

Since Lemma 6.24 does not require S to be finite (see the concluding remarks in [Wolfowitz (1963)]), Theorem 6.23 and 6.25 can be generalized as (see also [Shen (2000)]):

Theorem 6.26 *If $M(k)$ are taken from a compact set whose finite products are SIA, then a Markov chain with $M(k)$ as transition matrices is weakly ergodic and $\lim_{n\to\infty} M(n)M(n-1)\cdots M(1)$ converges to a rank one matrix of the form $\mathbf{1}c^T$.*

In [Wolfowitz (1963)] it was shown that $n_W \leq 2^{n^2}$, where n is the order of the matrices. In [Paz (1965)] it was shown that $n_W \leq \frac{1}{2}\left(3^n - 2^{n+1} + 1\right)$ which is a tight bound:

Lemma 6.27 *Let S be a set of matrices whose finite products are SIA. Then a product of matrices in S of length $\frac{1}{2}\left(3^n - 2^{n+1} + 1\right)$ or larger is scrambling.*

If all matrices have positive diagonal elements, then the constant n_W can be reduced to be linear in n [Wu (2006c)]:

Lemma 6.28 *Let S be a set of SIA matrices with positive diagonal elements, then any product of matrices in S of length $n-1$ or larger is scrambling and any product of matrices in S of length $n(n-2)+1$ or larger is Markov.*

Proof: Let $A_i \in S$. For the first part of the result, let $B(m) = A_1 A_2 \ldots A_m$. For $i \neq j$, we need to show that $B(n-1)_{ik} \neq 0$ and $B(n-1)_{jk} \neq 0$ for some k. Let $C_i(m)$ be the children of vertex i in the graph of $B(m)$. First note that since A_i has positive diagonal elements, $i \in C_i(m)$ and the graph of $B(m)$ is a subgraph of the graph of $B(m+1)$. Suppose that $C_i(m)$ does not intersect $C_j(m)$. Then the root of the directed tree r in the interaction graph of A_{m+1} is either not in $C_i(m)$, or not in $C_j(m)$. Suppose r is not in $C_j(m)$. Since all vertices in $C_j(m)$ has a path to r in the graph of A_{m+1}, at least one vertex in $C_j(m)$ must have a child in the graph A_{m+1} outside of $C_j(m)$. Since $C_j(m+1)$ is $C_j(m)$ plus the children of $C_j(m)$ in the graph of A_{m+1}, this means that $C_j(m+1)$ is strictly larger than $C_j(m+1)$. Similarly, if r is not in $C_i(m)$, then $C_i(m+1)$ is strictly larger than $C_i(m+1)$. Since either $C_i(1)$ or $C_j(1)$ has at least two elements (recall that $i \in C_i(m)$), this means that $C_i(n-1)$ intersects $C_j(n-1)$, say $k \in C_i(n-1) \cap C_j(n-1)$ which is the k we are looking for.

For the second part of the result, Let $\mathcal{G}_i$ be the interaction graph of A_i. Since each $\mathcal{G}_i$ contains a root vertex of a spanning directed tree, it is clear that there are $n-1$ graphs among $\mathcal{G}_i$ with the same root vertex r. Let us denote the corresponding matrices as $A_{m_1}, \dots, A_{m_{n-1}}$. Since A_i has positive diagonal elements, it suffices to show that the matrix product $A_{m_1} \cdots A_{m_{n-1}}$ is a Markov matrix.

Let us denote the children of r in the interaction graph of $A_{m_1} \cdots A_{m_i}$ as C_i. Note that C_1 has at least two elements. Since the diagonal elements of A_i are positive, C_{i+1} is equal to C_i plus the children of C_i in $\mathcal{G}_{m_{i+1}}$. Since $r \in C_i$ and is the root of a spanning directed tree in $\mathcal{G}_{m_{i+1}}$, the children of C_i in $\mathcal{G}_{m_{i+1}}$, must include some vertex not in C_i (unless $C_i = V$). This implies that $C_{n-1} = V$ which implies that the r-th column of $A_{m_1} \dots A_{m_{n-1}}$ is positive, i.e. A is Markov. □

In the first part of Lemma 6.28, a product of matrices A_i of length $n-1$ results in a scrambling matrix. Using the example

$$A_i = \begin{pmatrix} 1 & 1 & & & \\ & 1 & 1 & & \\ & & 1 & \ddots & \\ & & & \ddots & \ddots \\ & & & & 1 \end{pmatrix}$$

for all i, we see that the bound $n-1$ is tight in Lemma 6.28.

If the positive diagonal elements condition is omitted, then Lemma 6.28 is not true as the following example illustrates. For $M(k) = \begin{pmatrix} 0 & 1 \\ 1 & 0 \end{pmatrix}$, the products of $M(k)$ oscillate between $\begin{pmatrix} 0 & 1 \\ 1 & 0 \end{pmatrix}$ and $\begin{pmatrix} 1 & 0 \\ 0 & 1 \end{pmatrix}$.

Stochastic matrices with positive diagonal elements satisfy the Wolfowitz conditions in Theorem 6.23:

Lemma 6.29

(1) A stochastic matrix with positive diagonal elements is SIA if and only if it is irreducible or 1-reducible.
(2) Let A_i be stochastic matrices with positive diagonal elements. Then $\frac{1}{n}\sum_{i=1}^{n} A_i$ is SIA if and only if $A_1A_2\cdots A_n$ is SIA. In particular, if one of the A_i is SIA, then so is $A_1A_2\cdots A_n$.

Proof: By Corollary 2.27, if a stochastic matrix A with positive diagonal elements is irreducible or 1-reducible, then 1 is an isolated eigenvalue with all other eigenvalues having norm strictly less than 1. Thus A^k converges to a rank one matrix of the form $\mathbf{1}c^t$ as $k \to \infty$, i.e. A is SIA. On the other hand, if A is m-reducible for $m \geq 2$, then writing A in the form of Eq. (2.2) shows that A^n will have the blocks $B_{k+1}^n, \ldots, B_{k+m}^n$ on the diagonal. Since $B_{k+1}, \ldots, B_{k+m}$ are stochastic irreducible matrices, $B_{k+1}^n, \ldots, B_{k+m}^n$ do not vanish as $n \to \infty$ and thus A^n does not converge to a rank-one matrix. This proves the first statement. A_i's having positive diagonal elements implies that if the (k,l)-th element of A_j is nonzero for some j, then so is the corresponding element in $A_1A_2\cdots A_n$. This means that if the interaction graph of $\frac{1}{n}\sum_{i=1}^{n} A_i$ contains a spanning directed tree, then so does the interaction graph of $A_1A_2\cdots A_n$. On the other hand, suppose the interaction graph of $A_1A_2\cdots A_n$ contains a spanning directed tree. By Theorem 2.1 a graph containing a spanning directed tree is equivalent to the property of quasi-strongly connected. For each edge (i,j) in the graph of $A_1A_2\cdots A_n$ which is not in the graph of $\frac{1}{n}\sum_{i=1}^{n} A_i$, there is a directed path from i to j in the graph of $\frac{1}{n}\sum_{i=1}^{n} A_i$. Therefore the interaction graph of $\frac{1}{n}\sum_{i=1}^{n} A_i$ is also quasi-strongly connected. This proves the second statement. □

Corollary 6.30 *For a stochastic matrix with positive diagonal elements, A is SIA if and only if the interaction graph of A contains a spanning directed tree.*

Proof: Follows from Corollary 2.25, Corollary 2.27 and Lemma 6.29. □

6.5 Contractive matrices

Recently, synchronization of the state equation $x(k+1) = M(k)x(k)$ has been analyzed through the theory of paracontractive and pseudocontractive matrices [Xiao *et al.* (2005); Fang and Antsaklis (2005)]. In this section we show that there is a close relationship between this approach and the Markov chain approach in Section 6.4. The discussion here is based on [Wu (2006b)].

We start first with a definition of a paracontracting matrix [Nelson and Neumann (1987)]:

Definition 6.31 Let $\|\cdot\|$ be a vector norm in $\mathbb{C}^n$. An n by n matrix B is *nonexpansive* with respect to $\|\cdot\|$ if

$$\forall x \in \mathbb{C}^n, \|Bx\| \leq \|x\| \tag{6.19}$$

The matrix B is called *paracontracting* with respect to $\|\cdot\|$ if

$$\forall x \in \mathbb{C}^n, Bx \neq x \Leftrightarrow \|Bx\| < \|x\| \tag{6.20}$$

It is easy to see that normal matrices with eigenvalues in the unit circle and for which 1 is the only eigenvalue of unit norm is paracontractive with respect to $\|\cdot\|_2$.

Definition 6.32 For a vector $x \in \mathbb{C}^n$ and a closed set X^*, y^* is called a *projection vector* of x onto X^* if $y^* \in X^*$ and

$$\|x - y^*\| = \min_{y \in X^*} \|x - y\|$$

The distance of x to X^* is defined as $d(x, X^*) = \|x - P(x)\|$ where $P(x)$ is a projection vector of x onto X^*.

Even though the projection vector is not necessarily unique, we write $P(x)$ when it is clear which projection vector we mean or when the choice is immaterial.

Lemma 6.33 *If $x \in \mathbb{R}^n$ and $X^* = \{\alpha\mathbf{1} : \alpha \in \mathbb{R}\}$, the projection vector $P(x)$ of x onto X^* is $\alpha\mathbf{1}$ where:*

- *for the norm $\|\cdot\|_2$, $\alpha = \frac{1}{n}\sum_i x_i$ and $d(x, X^*) = \sqrt{\sum_i (x_i - \alpha)^2}$.*
- *for the norm $\|\cdot\|_\infty$, $\alpha = \frac{1}{2}(\max_i x_i + \min_i x_i)$, and $d(x, X^*) = \frac{1}{2}(\max_i x_i - \min_i x_i)$.*
- *for the norm $\|\cdot\|_1$, $d(x, X^*) = \sum_{i=\lceil \frac{n}{2} \rceil + 1}^{n} \hat{x}_i - \sum_{i=1}^{\lfloor \frac{n}{2} \rfloor} \hat{x}_i$ and*
  - *for n odd, $\alpha = \hat{x}_{\lceil \frac{n}{2} \rceil}$.*
  - *for n even, α can be chosen to be any number in the interval $[\hat{x}_{\frac{n}{2}}, \hat{x}_{\frac{n}{2}+1}]$.*

Here $\hat{x}_i$ are the values x_i rearranged in nondecreasing order $\hat{x}_1 \leq \hat{x}_2 \leq \cdots$.

Proof: The proof is left as an exercise for the reader. □

The property of paracontractivity is used to show convergence of infinite products of paracontractive matrices and this in turn is used to prove convergence in various parallel and asynchronous iteration methods [Bru *et al.* (1994)]. In [Su and Bhaya (2001)] this property is generalized to pseudocontractivity.

Definition 6.34 Let T be an operator on $\mathbb{R}^n$. The operator T is *nonexpansive* with respect to $\|\cdot\|$ and a closed set X^* if

$$\forall x \in \mathbb{R}^n, x^* \in X^*, \|Tx - x^*\| \leq \|x - x^*\| \tag{6.21}$$

T is *pseudocontractive* with respect to $\|\cdot\|$ and X^* if it is nonexpansive with respect to $\|\cdot\|$ and X^* and

$$\forall x \notin X^*, d(Tx, X^*) < d(x, X^*) \tag{6.22}$$

In [Su and Bhaya (2001)] in was shown that there are pseudocontractive nonnegative matrices which are not paracontractive with respect to $\|\cdot\|_\infty$ and proves a result on the convergence of infinite products of pseudocontractive matrices. Furthermore, [Su and Bhaya (2001)] studies a class of matrices for which a finite product of matrices from this class of length at least $n-1$ is pseudocontractive in $\|\cdot\|_\infty$.

We concentrate on the case where T are matrices and X^* is the span of the corresponding Perron eigenvector. If the Perron eigenvector is strictly positive, then as in [Su and Bhaya (2001)], a scaling operation $T \to W^{-1}TW$ where W is the diagonal matrix with the Perron eigenvector on the diagonal, transforms T into a matrix for which the Perron eigenvector is $\mathbf{1}$. Therefore in the sequel we will focus on constant row sum matrices with $X^* = \{\alpha\mathbf{1} : \alpha \in \mathbb{R}\}$.

6.5.1 *Pseudocontractivity and scrambling stochastic matrices*

A priori it is not clear if pseudocontractivity of a matrix can be easily determined. The following result shows that pseudocontractivity of stochastic matrices with respect to $\|\cdot\|_\infty$ is equivalent to the scrambling condition and thus can be easily determined.

Theorem 6.35 *Let A be a stochastic matrix. The matrix A is pseudocontractive with respect to $\|\cdot\|_\infty$ and $X^* = \{\alpha\mathbf{1} : \alpha \in \mathbb{R}^n\}$ if and only if A is a scrambling matrix.*

Proof: Let $x^* \in X^*$. Then $Ax^* = x^*$ and thus $\|Ax - x^*\|_\infty = \|A(x - x^*)\|_\infty \leq \|x - x^*\|_\infty$. Thus all stochastic matrices are nonexpansive with respect to $\|\cdot\|_\infty$ and X^*. Suppose A is a scrambling matrix. Then $\chi(A) > 0$, and $\delta(A) < 1$ by Theorem 6.13. By Lemma 6.33 and Theorem 6.14, A is pseudocontractive. Suppose A is not a scrambling matrix. Then there exists i,j such that for each k, either $A_{ik} = 0$ or $A_{jk} = 0$. Define x as $x_k = 1$ if $A_{ik} > 0$ and $x_k = 0$ otherwise. Since A is stochastic, it does not have zero rows and thus there exists k' and k'' such that $A_{ik'} = 0$ and $A_{ik''} > 0$. This means that $x \notin X^*$. Let $y = Ax$. Then $y_i = 1$ and $y_j = 0$. This means that $\max_i y_i - \min_i y_i \geq 1 = \max_i x_i - \min_i x_i$, i.e. A is not pseudocontractive. □

In [Lubachevsky and Mitra (1986)] the convergence of a class of asynchronous iteration algorithms was shown by appealing to results about scrambling matrices. In [Su and Bhaya (2001)] this result is proved using the framework of pseudocontractions. Theorem 6.35 shows the close relationship between these two approaches.

6.5.2 *Set-nonexpansive and set-contractive operators*

Consider the stochastic matrix

$$A = \begin{pmatrix} 0.5 & 0 & 0.5 \\ 0.5 & 0.5 & 0 \\ 0.5 & 0 & 0.5 \end{pmatrix}$$

The matrix A is not pseudocontractive with respect to the Euclidean norm $\|\cdot\|_2$ and $X^* = \{\alpha\mathbf{1} : \alpha \in \mathbb{R}\}$ since $\|A\|_2 = 1.088 > 1$. On the other hand, A satisfies Eq. (6.22)[2]. This motivates us to define the following generalization of pseudocontractivity:

Definition 6.36 Let X^* be a closed set in $\mathbb{R}^n$. An operator T on $\mathbb{R}^n$ is *set-nonexpansive* with respect to $\|\cdot\|$ and X^* if

$$\forall x \in \mathbb{R}^n, d(Tx, X^*) \leq d(x, X^*)$$

An operator T on $\mathbb{R}^n$ is *set-contractive* with respect to $\|\cdot\|$ and X^* if it is set-nonexpansive with respect to $\|\cdot\|$ and X^* and

$$\forall x \notin X^*, d(Tx, X^*) < d(x, X^*).$$

[2]This can be shown using Theorem 6.46.

The *set-contractivity* of an operator T is defined as

$$c(T) = \sup_{x \notin X^*} \frac{d(Tx, X^*)}{d(x, X^*)} \geq 0$$

There is a dynamical interpretation to Definition 6.36. If we consider the operator T as a discrete-time dynamical system, then T being set-nonexpansive and set-contractive imply that X^* is a globally nonrepelling invariant set and a globally attracting set of the dynamical system respectively [Wiggins (1990)].

Lemma 6.37 *The following statements are true:*

- *If T and S are linear operators, then $c(T+S) \leq c(T) + c(S)$.*
- *T is set-nonexpansive with respect to $\|\cdot\|$ and X^* if and only if $T(X^*) \subseteq X^*$ and $c(T) \leq 1$.*
- *If T is set-contractive with respect to $\|\cdot\|$ and X^*, then the fixed points of T is a subset of X^*.*
- *If $T_1(X^*) \subseteq X^*$, then $c(T_1 \circ T_2) \leq c(T_1)c(T_2)$.*

Proof: First note that

$$\begin{aligned} d(Tx + Sx, X^*) &= \|Tx + Sx - P(Tx + Sx)\| \\ &\leq \|Tx + Sx - P(Tx) - P(Sx)\| \\ &\leq \|Tx - P(Tx)\| + \|Sx - P(Sx)\| \\ &= d(Tx, X^*) + d(Sx, X^*) \end{aligned}$$

Then for linear T and S,

$$\begin{aligned} c(T+S) &= \sup_{x \notin X^*} \frac{d(Tx + Sx, X^*)}{d(x, X^*)} \leq \sup_{x \notin X^*} \frac{d(Tx, X^*)}{d(x, X^*)} + \sup_{x \notin X^*} \frac{d(Sx, X^*)}{d(x, X^*)} \\ &= c(T) + c(S) \end{aligned}$$

The second statement is true by definition. The proof of the third statement is the same as in Proposition 2.1 in [Su and Bhaya (2001)] As for the fourth statement, suppose $T_1(X^*) \subseteq X^*$. Let $x \notin X^*$. If $T_2(x) \in X^*$, then $d(T_1 \circ T_2(x), X^*) = 0$. If $T_2(x) \notin X^*$, then $d(T_1 \circ T_2(x)) \leq c(T_1) d(T_2(x), X^*) \leq c(T_1)c(T_2)d(x, X^*)$. □

Definition 6.38 a real-valued function f is a submultiplicative pseudonorm if

(1) $f(x) \geq 0$
(2) $f(0) = 0$

(3) $f(ax) = |a|f(x)$ for all scalars a
(4) $f(x+y) \leq f(x) + f(y)$
(5) $f(xy) \leq f(x)f(y)$

Lemma 6.39 *Let X^* be a closed set such that $\alpha X^* \subseteq X^*$ for all $\alpha \in \mathbb{R}$. Then c is a submultiplicative pseudonorm on the set of linear operators T such that $T(X^*) \subseteq X^*$.*

Proof: By Lemma 6.37 we only need to show that $c(aT) = |a|c(T)$. First note that $P(aTx) = aP(Tx)$. This implies that

$$d(aTx, X^*) = \|aTx - P(aTx)\| = \|aTx - aP(Tx)\| = |a|d(Tx, X^*)$$

□

Lemma 6.40 *Let X^* be a closed set such that $\alpha X^* \subseteq X^*$ for all $\alpha \in \mathbb{R}$. If T is linear and $T(X^*) \subseteq X^*$, then $c(T) = \sup_{\|x\|=1, P(x)=0} d(T(x), X^*)$.*

Proof: Let $\epsilon = \sup_{\|x\|=1, P(x)=0} d(T(x), X^*)$. Clearly $\epsilon \leq c(T)$. For $x \notin X^*$, 0 is a projection vector of $x - P(x)$. Since $T(P(x)) \in X^*$, this implies that $d(T(x), X^*) = d(T(x - P(x)), X^*) \leq \epsilon\|x - P(x)\| = \epsilon d(x, X^*)$, i.e. $\epsilon \geq c(T)$. □

Lemma 6.41 *Let X^* be a closed set such that $\alpha X^* \subseteq X^*$ for all $\alpha \in \mathbb{R}$. An set-nonexpansive matrix T is set-contractive with respect to X^* if and only if $c(T) < 1$.*

Proof: One direction is clear. Suppose T is set-contractive. By compactness

$$\sup_{\|x\|=1, P(x)=0} d(T(x), X^*) = \epsilon < 1$$

and the conclusion follows from Lemma 6.37 and Lemma 6.40. □

If T is nonexpansive with respect to $\|\cdot\|$ and X^*, then

$$\|Tx - P(Tx)\| \leq \|Tx - P(x)\| \leq \|x - P(x)\|$$

and T is set-nonexpansive. Thus set-contractivity is more general than pseudocontractivity. However, they are equivalent for stochastic matrices with respect to $\|\cdot\|_\infty$ and $X^* = \{\alpha \mathbf{1} : \alpha \in \mathbb{R}\}$.

Lemma 6.42 *With respect to $\|\cdot\|_\infty$ and $X^* = \{\alpha \mathbf{1} : \alpha \in \mathbb{R}\}$, a stochastic matrix T is pseudocontractive if and only if it is set-contractive.*

Proof: Follows from the fact that a stochastic matrix is nonexpansive with respect to $\|\cdot\|_\infty$ and $X^* = \{\alpha \mathbf{1} : \alpha \in \mathbb{R}\}$. □

Consider the following definition of a monotone norm [Horn and Johnson (1985)]:

Definition 6.43 A vector norm $\|\cdot\|$ on $\mathbb{R}^n$ is *monotone* if

$$\left\|(x_1,\cdots,x_n)^T\right\| \leq \left\|(y_1,\cdots,y_n)^T\right\|$$

for all x_i and y_i such that $|x_i| \leq |y_i|$. A vector norm $\|\cdot\|$ on $\mathbb{R}^n$ is *weakly monotone* if

$$\left\|(x_1,\cdots,x_{k-1},0,x_{k+1},\cdots,x_n)^T\right\| \leq \left\|(x_1,\cdots,x_{k-1},x_k,x_{k+1},\cdots,x_n)^T\right\|$$

for all x_i and k.

The next result gives a necessary condition of set-contractivity of a matrix in terms of its graph.

Theorem 6.44 *Let A be a constant row sum matrix with row sums r such that $|r| \geq 1$. If A is set-contractive with respect to a weakly monotone vector norm $\|\cdot\|$ and $X^* = \{\alpha\mathbf{1} : \alpha \in \mathbb{R}\}$, then the interaction digraph of A contains a spanning directed tree.*

Proof: If the interaction digraph A does not have a spanning directed tree, then the discussion in Section 2.2 shows that after simultaneous row and column permutation, A can be written as a block upper triangular matrix:

$$A = \begin{pmatrix} * & * & * & * \\ & \ddots & * & * \\ & & A_1 & 0 \\ & & & A_2 \end{pmatrix}$$

where $*$ are arbitrary entries and A_1 and A_2 are m_1 by m_1 and m_2 by m_2 square irreducible matrices respectively. Define $x = (0,\ldots,0,-a_1e_1,a_2e_2)^T \notin X^*$, where e_1 and e_2 are vectors of all 1's of length m_1 and m_2 respectively. Let $z = (0,\ldots,0,e_3)^T$ where e_3 is the vector of all 1's of length m_1+m_2 and $Z^* = \{\alpha z : \alpha \in \mathbb{R}\}$. Note that the set of projection vectors of a fixed vector x to Z^* is a convex connected set. Let αz be a projection vector of x to Z^*. Suppose that for $a_1 = a_2 \neq 0$, $\alpha \neq 0$. Since $-\alpha z$ is a projection vector of $-x$ to Z^* and α (or at least a choice of α) depends continuously on a_1 and a_2, by varying a_1 to $-a_1$ and varying a_2 to $-a_2$, α changes to $-\alpha$. This means that we can find a_1 and a_2 not both zero, such that 0 is a projection vector of x to Z^*. In this case

$x \notin X^*$ and by weak monotonicity $d(x, Z^*) = d(x, X^*) = \|x\|$. It is clear that $y = Ax$ can be written as

$$y = \begin{pmatrix} * \\ \vdots \\ * \\ -ra_1e_1 \\ ra_2e_2 \end{pmatrix}$$

Let βe be a projection vector of y onto X^*. By the weak monotonicity of the norm,

$$d(y, X^*) = \|y - \beta e\| \geq \begin{pmatrix} 0 \\ \vdots \\ 0 \\ (-ra_1 - \beta)e_1 \\ (ra_2 - \beta)e_2 \end{pmatrix} = r\left(x - \frac{\beta}{r}z\right)$$

Since 0 is a projection vector of x onto Z^*

$$d(y, X^*) \geq |r| d(x, Z^*) \geq d(x, X^*)$$

Thus A is not set-contractive. □

6.5.3 *Set-contractivity under the max-norm*

The following result shows that under the max-norm, the set-contractivity is equivalent to the quantity $\delta(A)$ used in the study of inhomogeneous Markov chains in Section 6.4.

Theorem 6.45 *Let A be a matrix with constant row sum r. Then $c(A) = r - \chi(A)$ with respect to $\|\cdot\|_\infty$ and $X^* = \{\alpha\mathbf{1} : \alpha \in \mathbb{R}\}$. In particular, the matrix A is set-nonexpanding with respect to $\|\cdot\|_\infty$ and $X^* = \{\alpha\mathbf{1} : \alpha \in \mathbb{R}\}$ if and only if $r - \chi(A) \leq 1$. The matrix A is set-contractive with respect to $\|\cdot\|_\infty$ and $X^* = \{\alpha\mathbf{1} : \alpha \in \mathbb{R}\}$ if and only if $r - \chi(A) < 1$.*

Proof: $c(A) \leq r - \chi(A)$ follows from Lemma 6.33, Theorem 6.13 and Theorem 6.14. Since $c(A) \geq 0$, $c(A) = r - \chi(A)$ if $r - \chi(A) = 0$. Therefore we assume that $r - \chi(A) > 0$. Let j and k be such that $\chi(A) = \sum_i \min(A_{ji}, A_{ki})$. Define x such that $x_i = 1$ if $A_{ji} < A_{ki}$ and $x_i = 0$ otherwise. Since $r - \chi(A) > 0$, x is not all 0's or all 1's, i.e. $x \notin X^*$. Let $y = Ax$. Then by

Lemma 6.33

$$\begin{aligned}2d(y, X^*) \geq y_k - y_i &= \sum_{i, A_{ji} < A_{ki}} A_{ki} - A_{ji} \\ &= \sum_i A_{ki} - \sum_{i, A_{ji} \geq A_{ki}} A_{ki} - \sum_{i, A_{ji} < A_{ki}} A_{ji} \\ &= r - \chi(A)\end{aligned}$$

Since $2d(x, X^*) = 1$, it follows that $c(A) \geq r - \chi(A)$. □

6.5.4 *Set-contractivity under the Euclidean norm*

The following result characterizes set-contractivity of matrices with respect to $\|\cdot\|_2$ in terms of matrix norms.

Theorem 6.46 *Let A be an n by n constant row sum matrix and K be an n by $n-1$ matrix whose columns form a orthonormal basis of $e^\perp$. Then $c(A) = \|AK\|_2$ with respect to $\|\cdot\|_2$ and $X^* = \{\alpha \mathbf{1} : \alpha \in \mathbb{R}\}$. In particular $\|AK\|_2 \leq 1$ if and only if A is set-nonexpanding with respect to $\|\cdot\|_2$ and $X^* = \{\alpha \mathbf{1} : \alpha \in \mathbb{R}\}$. Similarly, $\|AK\|_2 < 1$ if and only if A is set-contracting with respect to $\|\cdot\|_2$ and $X^* = \{\alpha \mathbf{1} : \alpha \in \mathbb{R}\}$.*

Proof: Define $J = \mathbf{1}\mathbf{1}^T$ as the n by n matrix of all $1's$. Note that $\|x\|_2 = \|Kx\|_2$ and $JK = 0$. Let $B = A - \frac{1}{n}J$. Then

$$\|AK\|_2 = \|BK\|_2 = \max_{\|x\|_2=1} \|BKx\|_2 = \max_{\|Kx\|_2=1} \|BKx\|_2 = \max_{x \perp e, \|x\|_2=1} \|Bx\|_2$$

By Lemma 6.33 $P(x) = \frac{1}{n}Jx$ and $d(Ax, X^*) = \|Bx\|$. Since A has constant row sums, $A(X^*) \subseteq X^*$ and by Lemma 6.40 $c(A) = \max_{P(x)=0, \|x\|_2=1} d(Ax, X^*) = \max_{P(x)=0, \|x\|_2=1} \|Bx\|_2$. Since $P(x) = 0$ if and only if $x \perp \mathbf{1}$, this means that $c(A) = \|AK\|_2$. □

6.5.5 *Set-contractivity under a weighted Euclidean norm*

Definition 6.47 Given a positive vector w, the weighted 2-norm $\|\cdot\|_w$ is defined as

$$\|x\|_w = \sqrt{\sum_i w_i x_i^2}$$

Theorem 6.48 *Let A be an n by n constant row sum matrix and K be as defined in Theorem 6.46. Let w be a positive vector such that $\|w\|_\infty = 1$ and $W = diag(w)$. Then $c(A) \leq \left\| W^{\frac{1}{2}} A W^{-1} K \right\|_2$ with respect to $\|\cdot\|_w$ and $X^* = \{\alpha \mathbf{1} : \alpha \in \mathbb{R}\}$.*

Proof: The proof is similar to Theorem 6.46. Define $J_w = \frac{\mathbf{1}w^T}{\sum_i w_i}$ and $B = A - J_w$. Note that $J_w W^{-1} K = 0$. Then

$$\begin{aligned}\|W^{\frac{1}{2}}AW^{-1}K\|_2 &= \|W^{\frac{1}{2}}BW^{-1}K\|_2 \\ &= \max_{\|Kx\|_2=1} \|W^{\frac{1}{2}}BW^{-1}Kx\|_2 \\ &= \max_{x\perp\mathbf{1},\|x\|_2=1} \|W^{\frac{1}{2}}BW^{-1}x\|_2\end{aligned}$$

Now $x\perp\mathbf{1}$ if and only if $W^{-1}x\perp w$. Since $\|x\|_2 = \|W^{-\frac{1}{2}}x\|_w$, this means that $\|W^{\frac{1}{2}}AW^{-1}K\|_2 = \max_{x\perp w,\|W^{\frac{1}{2}}x\|_w=1} \|W^{\frac{1}{2}}Bx\|_2$. Since $\|w\|_\infty = 1$, this means that $\|W^{\frac{1}{2}}x\|_w = \sqrt{\sum_i (w_i x_i)^2} \leq \|x\|_w$ and thus

$$\|W^{\frac{1}{2}}AW^{-1}K\|_2 \geq \max_{x\perp w,\|x\|_w=1} \|W^{\frac{1}{2}}Bx\|_2$$

It is straightforward to show that $P(x) = J_w x$ and thus $d(Ax, X^*) = \|Bx\|_w = \|W^{\frac{1}{2}}Bx\|_2$. Since A has constant row sums, $A(X^*) \subseteq X^*$ and by Lemma 6.40 $c(A) = \max_{P(x)=0,\|x\|_w=1} d(Ax, X^*) = \max_{P(x)=0,\|x\|_w=1} \left\|W^{\frac{1}{2}}Bx\right\|_2$. Since $P(x) = 0$ if and only if $x\perp w$, this means that $c(A) \leq \left\|W^{\frac{1}{2}}AW^{-1}K\right\|_2$. □

Note that the matrix A in Theorem 6.45, Theorem 6.46 and Theorem 6.48 is not necessarily nonnegative nor stochastic. Let us consider some matrices and look at their set-contractivities. The matrix

$$A_1 = \begin{pmatrix} 1.1 & 0.0 & 0.0 \\ 0.6 & 0.5 & 0 \\ 0.6 & 0 & 0.5 \end{pmatrix}$$

is set-contracting with respect to $\|\cdot\|_\infty$ and $X^* = \{\alpha\mathbf{1} : \alpha \in \mathbb{R}\}$ since $\chi(A_1) = 0.6$ and $c(A_1) = 1.1 - \chi(A_1) = 0.5 < 1$. It is not pseudocontracting with respect to $\|\cdot\|_\infty$ and $X^* = \{\alpha\mathbf{1} : \alpha \in \mathbb{R}\}$ since $\|A_1\|_\infty = 1.1 > 1$.

The stochastic matrix

$$A_2 = \begin{pmatrix} 0.4 & 0.3 & 0.3 \\ 0 & 1 & 0 \\ 0 & 0 & 1 \end{pmatrix}$$

is set-nonexpanding with respect to $\|\cdot\|_2$ and $X^* = \{\alpha\mathbf{1} : \alpha \in \mathbb{R}\}$ since $\|A_2K\|_2 = 1$ but it is not nonexpanding with respect to $\|\cdot\|_2$ and $X^* = \{\alpha\mathbf{1} : \alpha \in \mathbb{R}\}$ since $\|A_2\|_2 > 1$. Furthermore, Theorem 6.44 shows that A_2 is not set-contractive with respect to any weakly monotone norm and X^*.

The stochastic matrix

$$A_3 = \begin{pmatrix} 1 & 0 & 0 \\ 0.5 & 0.5 & 0 \\ 0 & 0.5 & 0.5 \end{pmatrix}$$

is set-contractive with respect to $\|\cdot\|_2$ and $X^* = \{\alpha\mathbf{1} : \alpha \in \mathbb{R}\}$ since $\|A_3K\|_2 = 0.939$. Since $\|A_3\|_2 > 1$ it is not nonexpanding nor pseudo-contractive with respect to $\|\cdot\|_2$ and X^*. It is also not pseudocontractive with respect to $\|\cdot\|_\infty$ and X^* since it is not scrambling.

The stochastic matrix

$$A_4 = \begin{pmatrix} 1 & 0 & 0 \\ 0.9 & 0.1 & 0 \\ 0.1 & 0.1 & 0.8 \end{pmatrix}$$

has an interaction digraph that contains a spanning directed tree. However, it is not set-nonexpanding with respect to $\|\cdot\|_2$ and $X^* = \{\alpha\mathbf{1} : \alpha \in \mathbb{R}\}$ since $\|A_4K\|_2 = 1.125 > 1$. This shows that the converse of Theorem 6.44 is not true for $\|\cdot\|_2$.[3] On the other hand, A_4 is set-contractive with respect to $\|\cdot\|_\infty$ and X^* since A_4 is a scrambling matrix. Furthermore, A_4 is set-contractive with respect to $\|\cdot\|_w$ and X^* for $w = (1, 0.2265, 1)^T$ since $\|W^{\frac{1}{2}}A_4W^{-1}K\|_2 < 1$.

Next we show some convergence results for dynamical systems of the form $x(k+1) = M(k)x(k)$ where some $M(k)$'s are set-contractive operators.

Theorem 6.49 *Let $\{M(k)\}$ be a sequence of set-nonexpansive operators with respect to $\|\cdot\|$ and X^* and suppose that*

$$\lim_{k\to\infty} \prod_k c(M(k)) = 0$$

Let $x(k+1) = M(k)x(k)$. For any initial vector $x(0)$, $\lim_{k\to\infty} d(x(k), X^*) = 0$.

Proof: From Lemma 6.37, $c(\prod_k M(k)) \le \prod_k c(M(k)) \to 0$ as $k \to \infty$ and the conclusion follows. □

Theorem 6.50 *Let $X^* = \{\alpha\mathbf{1} : \alpha \in \mathbb{R}\}$ and $\{M(k)\}$ be a sequence of n by n constant row sum nonnegative matrices such that*

- *the diagonal elements are positive;*

[3] Theorem 6.35 shows that the converse of Theorem 6.44 is false as well for stochastic matrices with respect to $\|\cdot\|_\infty$ and X^*.

- *all nonzero elements are equal to or larger than ϵ;*
- *the row sum is equal to or less than r.*

If $r^{n-1} - \epsilon^{n-1} < 1$ and for each k, the interaction digraph of $M(k)$ contains a spanning directed tree, then $\lim_{k\to\infty} d(x(k), X^*) = 0$ for the state equation $x(k+1) = M(k)x(k)$.

Proof: By Lemma 6.28 products of $n-1$ matrices $M(k)$ is scrambling. By definition, since each $M(k)$ has nonzero elements equal to or larger than ϵ, the nonzero elements of this product, denoted as P, will be equal to or larger than ϵ^{n-1}. This means that $\chi(P) \geq \epsilon^{n-1}$ and thus $\delta(P) \leq r^{n-1} - \epsilon^{n-1} < 1$ since P has row sums $\leq r^{n-1}$. Therefore P is set-contractive with respect to $\|\cdot\|_\infty$ and X^* with $c(P) \leq r^{n-1} - \epsilon^{n-1} < 1$. The result then follows from Theorem 6.49. □

The following result shows existence of linear operators $B(k)$ and vectors $x_k^* \in X^*$ such that $x(k+1) = B(k)x(k) + x_k^*$ has the same dynamics as $x(k+1) = M(k)x(k)$. In particular, for $y(k+1) = B(k)y(k)$ and $x(k+1) = M(k)x(k)$, $d(y(k), X^*) = d(x(k), X^*)$ for all k.

Theorem 6.51 *T is a set-nonexpansive operator with respect to $\|\cdot\|_\infty$ and $X^* = \{\alpha\mathbf{1} : \alpha \in \mathbb{R}\}$ if and only if for each $x \in \mathbb{R}^n$ there exists a stochastic matrix B and a vector $x^* \in X^*$such that $T(x) = Bx + x^*$.*

T is a set-contractive operator with respect to the norm $\|\cdot\|_\infty$ and $X^* = \{\alpha\mathbf{1} : \alpha \in \mathbb{R}\}$ if and only if for each $x \in \mathbb{R}^n$ there exists a scrambling stochastic matrix B and a vector $x^* \in X^*$ such that $T(x) = Bx + x^*$.

Proof: One direction of both statements follows from Theorem 6.35. Suppose T is set-nonexpansive and fix $x \in \mathbb{R}^n$. Define $x^* = P(T(x)) - P(x)$ which is a vector in X^*. Let $y = T(x) - x^*$. Then $P(y) = P(T(x)) - x^* = P(x)$ and by Lemma 6.33,

$$\min_i x_i \leq \min_i y_i \leq \max_i y_i \leq \max_i x_i$$

and thus there exists a stochastic matrix B such that $Bx = y$.

If T is set-contractive, then for $x \in X^*$, we can choose $B = \frac{1}{n}\mathbf{1}\mathbf{1}^T$ and $T(x) - Bx \in X^*$. For $x \notin X^*$, $d(x, X^*) < d(T(x), X^*)$. Define x^* and y as before and we see that

$$\min_i x_i < \min_i y_i \leq \max_i y_i < \max_i x_i$$

If $x_{i'} = \min_i x$, then it is clear that we can pick B with $Bx = y$ such that the i'-th column of B is positive, i.e. B is scrambling. □

It can be beneficial to consider set-contractivity with respect to different norms. For instance, consider $x(k+1) = M(k)x(k)$ where $M(k)$ are matrices that are not pseudocontractive with respect to $\|\cdot\|_\infty$ and $X^* = \{\alpha\mathbf{1} : \alpha \in \mathbb{R}\}$ and whose diagonal elements are 0. Since the diagonal elements are not positive, the techniques in [Su and Bhaya (2001)] cannot be used to show that products of $M(k)$ are pseudocontractive with respect to $\|\cdot\|_\infty$ and $X^* = \{\alpha\mathbf{1} : \alpha \in \mathbb{R}\}$. However, it is possible that $M(k)$ are set-contractive with respect to a different norm and thus convergence of $x(k)$ can be obtained by studying set-contractivity using this norm. For instance, the stochastic matrix

$$A = \begin{pmatrix} 0 & 0.5 & 0.5 \\ 1 & 0 & 0 \\ 0.5 & 0.5 & 0 \end{pmatrix}$$

has zeros on the diagonal and is not pseudocontractive with respect to $\|\cdot\|_\infty$ and $X^* = \{\alpha\mathbf{1} : \alpha \in \mathbb{R}\}$ since A is not scrambling. On the other hand, A is set-contractive with respect to $\|\cdot\|_2$ and $X^* = \{\alpha\mathbf{1} : \alpha \in \mathbb{R}\}$ since $\|AK\|_2 = 0.939 < 1$.

For a set of constant row sum matrices $M(k)$ and $x(k+1) = M(k)x(k)$, a lower bound for the exponential rate at which $x(k)$ approach $X^* = \{\alpha\mathbf{1} : \alpha \in \mathbb{R}\}$ is $-\ln(c(A))$. The above examples show that there are matrices for which this rate is 0 for $\|\cdot\|_\infty$ and positive for $\|\cdot\|_2$ and other matrices for which the rate is positive and 0 for $\|\cdot\|_\infty$ and $\|\cdot\|_2$ respectively.

On the other hand, even though set-contractivity depends on the norm used, the equivalence of norms on $\mathbb{R}^n$ and Lemma 6.41 provides the following result.

Theorem 6.52 *Let X^* be a closed set such that $\alpha X^* \subseteq X^*$ for all $\alpha \in \mathbb{R}$. and let H be a compact set of set-contractive matrices with respect to $\|\cdot\|_p$ and X^*. Then there exists m such that a product of m matrices in H is set-contractive with respect to $\|\cdot\|_q$.*

Corollary 6.53 *Let H be a compact set of stochastic set-contractive matrices with respect to $\|\cdot\|_p$ and $X^* = \{\alpha\mathbf{1} : \alpha \in \mathbb{R}\}$. Then a sufficiently long product of matrices in H is scrambling.*

6.5.6 *Set-contractivity and coefficient of ergodicity*

We illustrate here the relationship between set-contractivity and ergodicity coefficient. Let us define H as the set of stochastic matrices that are set-nonexpansive with respect to a norm $\|\cdot\|$ and $X^* = \{\alpha\mathbf{1} : \alpha \in \mathbb{R}\}$. For $\|\cdot\|_\infty$, H is the set of stochastic matrices. Let us define $\chi_c(A) = 1 - c(A)$. Then χ_c is a proper coefficient of ergodicity (Definition 6.19) when restricted to H. This can be seen as follows. Clearly $0 \leq \chi_c(A) \leq 1$. If $A = \mathbf{1}v^T$, then $Ax \in X^*$ and thus $c(A) = 0$ and $\chi_c(A) = 1$. If $A \neq \mathbf{1}v^T$, then there exists i, j, k such that $A_{ik} \neq A_{jk}$. Let x be the k-th unit basis vector. Then $(Ax)_i \neq (Ax)_j$, i.e. $d(Ax, X^*) > 0$, $c(A) > 0$ and $\chi_c(A) < 1$. By choosing $\chi_1 = \chi_2 = \chi_c$, Eq. (6.17) is satisfied with $C = 1$ by Lemma 6.37. This together with Theorem 6.20 shows that a sufficient and necessary condition for a sequence of matrices in H to be weakly ergodic is

$$\sum_{j=1}^{\infty} 1 - c(A_{i_j+1} \cdots A_{i_{j+1}}) = \infty$$

for some strictly increasing subsequence $\{i_j\}$.

Similarly, we can apply the results in [Neumann and Schneider (1999)] to show that

Theorem 6.54 *A sequence of stochastic matrices A_i is weakly ergodic if $\sum_{i=1}^{\infty} \max(c(A_i) - 1, 0)$ converges and $\sum_{i=1}^{\infty} \max(1 - c(A_i), 0)$ diverges.*

6.6 Further reading

The reader is referred to [Seneta (1973, 1979)] for an excellent review and history of the development of ergodicity coefficients and their applications to analyzing inhomogeneous Markov chains.

Chapter 7

Agreement and Consensus Problems in Groups of Interacting Agents

Recently, there has been much interest in studying cooperative behavior in interacting agents each executing a local protocol. In this chapter we study protocols for these agents that result in a common collective behavior. In particular, we consider systems where the state equations can be written as a linear system. In this case, the results in Chapter 6 can be used to derive conditions for consensus and agreement. These models have applications in modeling flocking behavior in birds and in coordination of mobile robots with limited range wireless communication capabilities.

7.1 Continuous-time models

In [Olfati-Saber and Murray (2004)] consensus behavior in a group of autonomous agents is modelled via a set of linear differential equations. For the autonomous case, this is given by:

$$\dot{x} = -Lx \tag{7.1}$$

where $x \in \mathbb{R}^n$ and L is the Laplacian matrix of a graph with n vertices and the state of each agent x_i is a scalar. Synchronization of Eq. (7.1) is referred to in this context as solving an *agreement* or *consensus* problem.

Corollary 2.25 can be used to prove that:

Theorem 7.1 *The state x in Eq. (7.1) solves an agreement problem for all initial x if and only if the interaction graph of L contains a spanning directed tree.*

Similar to the result in Section 4.1, this result is intuitive since the existence of a spanning directed tree in the interaction graph implies that there is a root vertex which influences directly or indirectly all other vertices. If no such spanning directed tree exists, then there exists two groups of vertices which do not influence each other and thus cannot possibly reach an agreement for arbitrary initial disagreement. Since the agent at the root vertex of the spanning directed tree influences all other agents, it can be considered a leader, which in general is not unique if there are more than one spanning directed tree.

If the interaction graph of L contains a spanning directed tree, then by Corollary 2.25 the zero eigenvalue is simple. Let w be such that $w^T L = 0$ and $\|w\|_1 = 1$.

Lemma 7.2 *[Olfati-Saber and Murray (2004)]* $\lim_{t\to\infty} e^{-Lt} = \mathbf{1}w^T$.

Proof: Write $-L$ in Jordan normal form as $-L = SJS^{-1}$ and thus $e^{-Lt} = Se^{Jt}S^{-1}$. Since all nonzero eigenvalues of $-L$ has negative real parts (Corollary 2.25), as $t \to \infty$, e^{Jt} converges (after possible simultaneous row and column permutation) to a diagonal matrix Q with a single nonzero element $Q_{11} = 1$. The first column and row of S and S^{-1} correspond to the right and left eigenvector of $-L$ for the zero eigenvalues, i.e. $\mathbf{1}$ and w^T respectively. This implies that $\lim_{t\to\infty} e^{-Lt} = SQS^{-1} = \mathbf{1}w^T$. □

Lemma 7.2 says that for Eq. (7.1) $x(t) \to (\sum_i w_i x_i(0))\mathbf{1}$ and the consensus state is a convex combination of the initial states of all the agents. If the i-th vertex does not have a directed path to the j-vertex in the interaction graph of L, then the consensus state reached by x_j cannot depend on $x_i(0)$. This implies that $w_i = 0$. This fact also follows from Lemma 2.29.

Definition 7.3 If $x \to x^*$ with $x_i^* = x_j^* = \frac{1}{n}\sum_i x_i(0)$, then Eq. (7.1) is said to solve the *average consensus* problem.

Theorem 7.4 *[Wu (2005a)] Let L be the Laplacian matrix of a strongly connected graph and let w be a positive vector such that $w^T L = 0$. Let W be a diagonal matrix with w_i on the diagonal. Then $W^{-1}\dot{x} = -Lx$ solves the average consensus problem.*

Proof: Since $\mathbf{1}^T W = w$, $\mathbf{1}^T WL = 0$ and the discussion above shows that for $\dot{x} = -WLx$, $x(t) \to \sum_i x_i(0)\mathbf{1}$, i.e. $\dot{x} = -WLx$ solves the average consensus problem. □

The special case of balanced graphs, where $W = I$, was studied in [Olfati-Saber and Murray (2004)].

7.1.1 Rate of exponential convergence

We say that $x(t)$ converges exponentially towards $x^*(t)$ with rate k if $\|x(t)-x^*(t)\| \leq O(e^{-kt})$. Since Eq. (7.1) is linear and nonautonomous, clearly x converges towards x^* in Eq. (7.1) with rate at least[1] $\mu_2(L)$ (Definition 4.12) which is positive for interaction graphs with a spanning directed tree. Since $\mu_2(L) \geq a_3(L)$ by Theorem 4.13 and Theorem 4.19, we get the following result on the convergence rate with respect to the algebraic connectivity.

Theorem 7.5 *[Wu (2005a)] If the graph of L is strongly connected then Eq. (7.1) synchronizes with rate $a_3(L) > 0$.*

The special case of Theorem 7.5 for balanced graphs in Eq. (7.1) was shown directly in [Olfati-Saber and Murray (2004)] using a quadratic Lyapunov function.

When the graph is undirected, $\mu_2 = \lambda_2$ is the algebraic connectivity of the graph. Thus similar to the algebraic connectivity, when the graph is undirected, adding extra undirected edges cannot decrease μ_2 [Fiedler (1973); Wu (2003a)]. However, this is not true for digraphs, illustrated using the same examples as in Section 2.5.

For the directed path graph in Figure 2.9(e) with Laplacian matrix L, $\mu_2(L) = 1$ since it is a directed tree and is isomorphic to its reversal (Corollary 4.18). By adding one directed edge we get the directed cycle graph in Figure 2.9(f) with a circulant Laplacian matrix L and $\mu_2(L) = 1 - \cos\left(\frac{2\pi}{n}\right)$ (see e.g. [Wu (2002)]).

Thus similar to a_4, by adding a single edge, μ_2 changes from $\mu_2 = 1$ to $\mu_2 = 1 - \cos\left(\frac{2\pi}{n}\right)$ which decreases to 0 as $O\left(\frac{1}{n^2}\right)$. Again this can be explained via the strongly connected components of these graphs.

7.1.2 Dynamic coupling topology

In this case the coupling topology, expressed through L, can change with time and the state equations is

$$\dot{x} = -L(t)x$$

Using Corollary 4.23 with $V = I$ and $f = 0$, the following result can be shown:

Theorem 7.6 *[Wu (2005a)] Eq. (7.1) solves the agreement problem with rate* $\inf_t a_1(L(t))$.

[1] Or at least arbitrarily close to $\mu_2(L)$, if L has nontrivial Jordan blocks.

Proof: Using the transformation $y = xe^{kt}$, we obtain $\dot{y} = \dot{x}e^{kt} + ky = ky - L(t)y$. Applying Theorem 4.4 with $f(x_i, t) = kx_i$ and $V = D(t) = 1$ and choosing $-w > k$, we obtain $y_i \rightarrow y_j$ if there exists U such that $U(-L(t) + kI)$ is negative definite for all t. As $x = ye^{-kt}$ this implies that $x_i \rightarrow x_j$ with rate k. The proof of Theorem 4.8 shows that for $U = I - \frac{1}{n}\mathbf{1}\mathbf{1}^T$, $U(-L(t) + kI)$ is negative definite if $a_1(L(t)) > k$. □

This problem for the special case of strongly connected balanced graphs was studied in [Olfati-Saber and Murray (2004)].

7.2 Discrete-time models

In [Vicsek *et al.* (1995)] flocking behavior in a group of autonomous agents is modelled via a set of linear nonautonomous discrete-time evolution equations. Flocking behavior corresponds to all agents moving in unison. This is expressed as all agents acquiring the same heading. By representing the heading as a state vector, flocking is then equivalent to synchronization. The state equations can be described by:

$$x(k+1) = M(k)x(k) \tag{7.2}$$

The results concerning SIA matrices in Section 6.4 were first used by [Jadbabaie *et al.* (2003)] to derive conditions for consensus of Eq. (7.2) by studying sequences of matrices whose products satisfy the Wolfowitz conditions.

Lemma 7.7 *Consider two Markov chains with transition matrices $Q(i)$ and $P(i)$. Suppose there exists an increasing sequence of positive integers m_i such that $Q(k) = P(m_k)P(m_k - 1)\dots P(m_{k-1} + 1)$. Then one Markov chain is weakly ergodic if and only if the other Markov chain is weakly ergodic.*

Proof: One direction is clear. Suppose now that the Markov chain corresponding to $Q(k)$ is weakly ergodic. Let $\epsilon > 0$. Then for some k, $\delta(Q(k)\dots Q(1)) < \epsilon$. For $m > m_k$, $P(m)P(m-1)\dots P(1)$ can be written as $P(m)P(m-1)\dots P(m_k + 1)Q(k)\dots Q(1)$. Since by Lemma 6.15 $\delta(AB) \leq \delta(A)\delta(B)$, this implies that $\delta(P(m)P(m-1)\dots P(1)) \leq \epsilon$. [2] □

Definition 7.8 S_d is defined as the set of stochastic matrices with positive diagonal elements. $S_d(v)$ is defined as the matrices in S_d such that

[2] See [Shen (2000)] for another proof.

each nonzero element is larger than v.

Theorem 7.9 *Let $\mathcal{G}(k)$ be the weighted interaction digraph of $M(k)$. Suppose there exists $v > 0$, $N > 0$ and an infinite sequence $k_1 \leq k_2 \leq \cdots$ such that*

(1) $M(k) \in S_d(v)$ for all k,
(2) $k_{i+1} - k_i \leq N$,
(3) For each i, the union of the graphs $\mathcal{G}(k_i), \mathcal{G}(k_i + 1), \cdots, \mathcal{G}(k_{i+1} - 1)$ contains a spanning directed tree,

then $x \to x^*$ as $t \to \infty$ in Eq. (7.2), where $x_i^* = x_j^*$ for all i, j.

Proof: Without loss of generality, assume $v < 1$. The product $P_i = M_{k_{i+1}-1} \cdots M_{k_i+1} M_{k_i}$ is in $S_d(v^N)$ and P_i are SIA matrices whose products are SIA by Lemma 6.29 and Corollary 6.30. Lemmas 6.27 and 6.28 show that $B_i = P_{m(i+1)} \cdots P_{mi+2} P_{mi+1}$ is a scrambling matrix for some integer m. Since $\chi(B_i) > 0$ and $B_i \in S_d(v^{Nm})$, this means that $\chi(B_i) \geq v^{Nm} > 0$. By Lemma 6.8 and Theorem 6.17 $\lim_{n\to\infty} B_n \cdots B_0 = \mathbf{1}c^T$ and the result then follows from Lemma 7.7. □

The constant $N < \infty$ is important in Theorem 7.9. Consider the example in [Moreau (2003)] where the following 4 stochastic matrices are defined:

$$M_a = \begin{pmatrix} \frac{1}{2} & \frac{1}{2} & 0 \\ 0 & 1 & 0 \\ 0 & 0 & 1 \end{pmatrix}, M_b = \begin{pmatrix} \frac{1}{2} & \frac{1}{2} & 0 \\ \frac{1}{2} & \frac{1}{2} & 0 \\ 0 & 0 & 1 \end{pmatrix}$$

$$M_c = \begin{pmatrix} 1 & 0 & 0 \\ 0 & 1 & 0 \\ 0 & \frac{1}{2} & \frac{1}{2} \end{pmatrix}, M_d = \begin{pmatrix} 1 & 0 & 0 \\ 0 & \frac{1}{2} & \frac{1}{2} \\ 0 & \frac{1}{2} & \frac{1}{2} \end{pmatrix}$$

Construct the matrices B_i:

$$B_i = \underbrace{M_a, \cdots, M_a}_{2i}, M_b, \underbrace{M_c, \cdots, M_c}_{2i+1}, M_d$$

It was shown in [Moreau (2003)] that the sequence of matrices $M(k)$ formed by concatenating $B_0, B_1, \ldots$ results in a dynamical system Eq. (7.2) which does not synchronize. Note that the graphs of B_i are strongly connected. Thus even though the conditions (1) and (3) of Theorem 7.9 are satisfied, B_i consists of arbitrarily long sequence of matrices M_k and thus a fixed N does not exist to satisfied condition (2) and no consensus is reached among agents. In other words, it is not sufficient (although it is easy to see that it is necessary) in order to reach consensus to have two sequences k_i, n_i such that the union of $\mathcal{G}(k_i)$,$\mathcal{G}(k_i+1), \ldots, \mathcal{G}(k_i+n_i)$ contains a spanning directed tree for all i. Furthermore, a modification of this example shows that the hypothesis in Theorem 7.9 is sufficient, but not necessary for consensus.

On the other hand, if each digraph $\mathcal{G}(k)$ is a disjoint union of strongly connected components, we show next that the constant N is not necessary in Theorem 7.9, i.e. $k_{i+1} - k_i$ can be arbitrarily large.

Definition 7.10 Let M_{bd} denote the set of matrices which can be written (possibly after simultaneous row and column permutation) in block diagonal form, where each diagonal block is irreducible.

Equivalently, matrices in M_{bd} are those matrices whose digraph consists of disjoint strongly connected components. M_{bd} can also be characterized as the matrices whose digraph does not contain a subgraph which is weakly connected and not strongly connected. An important subclass of matrices in M_{bd} are the stochastic matrices A such that $A_{ij} > 0 \Leftrightarrow A_{ji} > 0$, i.e. the graph of A (after ignoring the weights on the edges) is undirected.

Lemma 7.11 *[Coppersmith and Wu (2005)] If $\delta \leq 1$ and $A_i \in S_d(\delta) \cap M_{bd}$ for all i, then $A_1 A_2 \cdots A_k \in S_d(\delta^{n-1})$ for any k. Here n is the order of the matrices A_i.*

Proof: Fix an index l. Let $z(k)$ be the l-th row of $A_1 A_2 \cdots A_k$. We show by induction on k that the sum of any j elements of $z(k)$ is either 0 or greater than or equal to δ^{n-j} for each $j = 1, \ldots, n$. This is clearly true for $k = 1$. Note that $z(k) = z(k-1)A_k$. Pick n_j elements of $z(k)$, say with index set $J \subset \{1, \ldots, n\}$. They corresponds to the matrix product of $z(k-1)$ with n_j columns of A_k. Let K be the indices of the rows of A_k which have nonzero intersection with some of these columns. Clearly $J \subset K$ since A_k has positive diagonal elements. If $K = J$, then $A_k(i,j) = 0$ for $i \notin J$, $j \in J$. Since $A_k \in M_{bd}$ this implies that $A_k(i,j) = 0$ for $i \in J, j \notin J$. This implies that the elements of $A_k(i,j)$ for $i, j \in J$ correspond to a n_j by n_j stochastic submatrix, and thus the sums of the n_j elements in $z(k)$ is the same as the sum of these corresponding elements in $z(k-1)$.

If K is strictly larger than J, then the sum of the n_j elements in $z(k)$ is larger than or equal to the sum of the elements in $z(k-1)$ with indices in K multiplied by nonzero elements in A_k each of which is at least δ. Since the sum of the elements in $z(k-1)$ with indices in K is at least δ^{n-j-1} or zero, the result follows. $\square$

Remark The bound δ^{n-j} is best possible in the proof of Lemma 7.11. Let A_i be the identity matrix except that $A_i(i,i) = A_i(i+1,i+1) = 1-\delta$, $A_i(i,i+1) = A(i+1,i) = \delta$. Then $A_1 A_2 \ldots A_{n-1}$ has an element of size δ^{n-1} and the j-th row of $A_1 A_2 \ldots A_{n-j}$ has j elements whose sum is δ^{n-j}. If we remove the condition of positive diagonal elements, then Lemma 7.11 is false. For instance, take

$$A_1 = \begin{pmatrix} 0 & \frac{1}{2} & \frac{1}{2} \\ 1 & 0 & 0 \\ 1 & 0 & 0 \end{pmatrix}, A_2 = \begin{pmatrix} 0 & 1 & 0 \\ \frac{1}{2} & 0 & \frac{1}{2} \\ 0 & 1 & 0 \end{pmatrix}$$

Then $(A_1 A_2)^m$ has elements of size $\frac{1}{4^m}$.

Theorem 7.12 *If there exists $\delta > 0$ and an infinite sequences $m_1 < m_2 < \cdots$ such that $A_i \in S_d(\delta) \cap M_{bd}$ and $A_{m_i} A_{m_i+1} \cdots A_{m_{i+1}-1}$ is SIA for all i then the Markov chain with transition matrices A_i is weakly ergodic and $A_n A_{n-1} \ldots A_1$ converges to a rank-one matrix $\mathbf{1}c^T$ as $n \to \infty$.*

Proof: Without loss of generality, we can assume $\delta \leq 1$. By Lemma 7.11, $P_i = A_{m_i} \cdots A_{m_{i+1}-1}$ are elements of $S_d(\delta^{n-1})$. By Lemma 6.24, $B_i = P_{t_i+1} P_{t_i+2} \cdots P_{t(i+1)} \in S(\delta^{t(n-1)})$ is scrambling. Since $\gamma(B_i) > 0$, this means that $\gamma(B_i) \geq \delta^{t(n-1)} > 0$. Since the lower bound $\delta^{t(n-1)}$ does not depend on the length of the products in B_i, the same proof as Theorem 7.9 can be used to prove the result. □

Note that there is no uniform bound on the length of the matrix product in Theorem 7.12, i.e., $m_{i+1} - m_i$ can be arbitrarily large. In particular, this means that the condition (2) is not necessary in Theorem 7.9 if $M(k) \in M_{bd}$. In other words, Theorem 7.12 can be described in terms of the graphs of A_i as follows: The Markov chain is weakly ergodic if for all i,

- $A_i \in S_d(\delta)$;
- The digraph of A_i consists of disjoint strongly connected components;
- The union of the reversal of the digraphs of $A_{m_i}, A_{m_i+1}, \cdots, A_{m_{i+1}-1}$ contains a spanning directed tree.

As mentioned before, if $\mathcal{G}(k)$ (after ignoring the weights on the edges) are undirected graphs, then $\mathcal{G}(k)$ are disjoint union of strongly connected components. This implies that there is no need for a uniform bound N in the results in [Jadbabaie *et al.* (2003)].

Suppose that some of the matrices $M(k)$ are stochastic matrices that are not SIA, while the rest satisfies Theorem 7.9, would we still have consensus? The answer is no, as the following example indicates. Consider the stochastic matrices:

$$A = \begin{pmatrix} 1 & 0 & 0 \\ 0.5 & 0.5 & 0 \\ 0 & 0.5 & 0.5 \end{pmatrix}, \quad B = \begin{pmatrix} 1 & 0 & 0 \\ 0 & 0 & 1 \\ 0 & 0 & 1 \end{pmatrix}$$

The matrix $A \in S_d$ is SIA and system (7.2) with $M(k) = A$ reaches consensus. However, setting $M(k) = A$ when k is even and $M(k) = B$ when k is odd will not result in consensus since BA is a decomposable matrix and decouples the agents from interacting with each other.

On the other hand, Theorem 6.17 and Lemmas 6.27 and 6.28 show that we can still have consensus if the matrices that are not SIA are sparse enough among $M(k)$.

7.2.1 *Follow the leader dynamics and leadership in coordinated agents*

Ref. [Jadbabaie *et al.* (2003)] also considered a follow-the-leader configuration, where n agents are connected via an undirected graph. An additional agent, the leader, influences some of these n agents, but is itself not influenced by other agents. In other words, the state of the leader is constant.

In general we can start with a strongly connected digraph and have an additional leader or root vertex influences some of the vertices. This is illustrated in Figure 7.1. Since the leader vertex has indegree 0, every spanning directed tree must have the leader vertex as root.

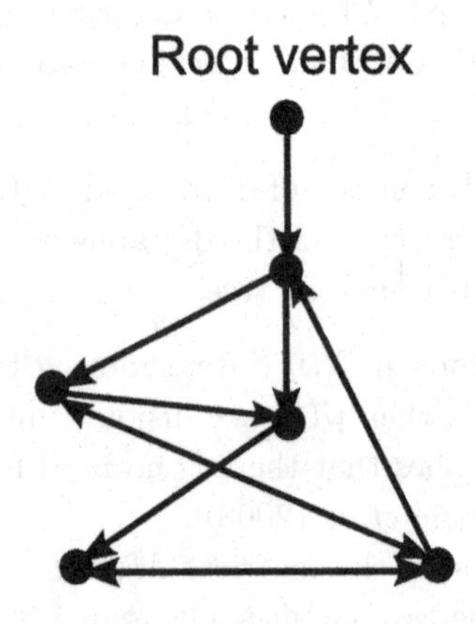

Fig. 7.1 A directed graph with a unique leader vertex.

We can generalize the concept of a leader as follows. A digraph can be partitioned into strongly connected components (SCC) using linear time algorithms [Tarjan (1972)]. From this structure we create a condensation directed graph (Definition 2.5). By Lemma 2.6 the condensation digraph does not contain directed cycles. Thus when a graph $\mathcal{G}$ contains a spanning directed tree, the unique strongly connected component which corresponds to the root of the condensation digraph can be considered a "leading" strongly connected component (LSCC), with the property that agents in the LSCC influencing all other agents outside the component, but not vice versa. When $\mathcal{G}$ changes with time, the LSCC also changes with time. It is clear that the roots of spanning directed trees are equal to the vertices in LSCC.

When $\mathcal{G}$ does not change with time, an alternative way of viewing the dynamics is the following. First the agents in the LSCC reach a consensus. Their states are then "collapsed" into a single "leader" state. The agents that the LSCC influences then reach a consensus following the "leader" state and are absorbed into the "leader" state etc, until finally all agents reach a consensus. This reduces the problem to the case of a single leader.

We can consider a range of "leadership" in the collection of agents, with the set of root vertices of spanning directed trees (i.e. the vertices in LSCC) as the leaders in the system. The system can be considered leaderless if the size of the size of the LSCC approaches the number of agents.

7.3 A nonlinear model for consensus

Recently, a nonlinear approach to consensus problem is proposed in [Moreau (2005)]. In this framework, the system is defined by the nonlinear equations $x_i(k+1) = f_i(x_1(k), x_2(k), \ldots x_n(k))$. For each time k, a directed graph $\mathcal{G}_k$ is associated with the dynamical system where each state vector x_i corresponds to a vertex and there is an edge from every vertex to itself. Let $N(i)$ be the set of the parents of vertex x_i in the directed graph. Note that $x_i \in N(i)$. The continuous functions f_i are defined such that $x_i(k+1)$ is in the convex hull of the $x_j(k)$'s with $x_j \in N(i)$. Furthermore, $x_i(k+1)$ is in the relative interior of the convex hull of the $x_j(k)$'s with $x_j \in N(i)$ when this relative interior is not empty. This system can be recast as Eq. (6.6). Since $x_i(k+1)$ is in the convex hull, it can be written as a convex combination of $x_1(k), x_2(k), \ldots, x_n(k)$, i.e. $x_i(k+1) = \sum_j m_{ij} x_j(k)$ where m_{ij} are nonnegative numbers such that $\sum_j m_{ij} = 1$. These numbers m_{ij}

form a stochastic matrix $M(k)$. In other words, at each time k a stochastic matrix $M(k)$ can be chosen such that $x(k+1) = (M(k) \otimes I)x(k)$, i.e. Eq. (6.6) for the case $D(k) = I$ and $u(k) = 0$.

Next we show that the relative interior condition implies that $M(k)$ can be chosen to have positive diagonal elements. If the relative interior is empty, the convex hull is a single point and thus $x_i(k) = x_j(k)$ for all $x_j \in N(i)$. This means that $x_i(k+1) = x_i(k)$, i.e. $m_{ii} = 1$. Otherwise, $x_i(k+1)$ is in the relative interior of the convex hull of $x_j(k)$, $x_j \in N(i)$. Consider the equation $x_i(k+1) = \sum_j m_{ij} x_j(k)$. Let $x_{j_1}, \cdots, x_{j_q}$ be a minimal set of extremal points that generate the convex hull. Then $x_i(k)$ and $x_i(k+1)$ can be written as $x_i(k) = \sum_{l=1}^{q} r_l x_{j_l}(k)$ and $x_i(k+1) = \sum_{l=1}^{q} h_l x_{j_l}(k)$ where the coefficients h_l are strictly positive. Since $x_i(k+1) = \sum_{l=1}^{q}(h_l - \alpha r_l)x_{j_l} + \alpha x_i(k)$, by choosing $\alpha > 0$ sufficiently small, we have written $x_i(k+1)$ as a convex combination such that the coefficient of $x_i(k)$ is strictly positive, i.e. $m_{ii} > 0$. In fact, this argument shows that $M(k)$ can be chosen such that $M_{ij}(k) > 0$ if and only if $x_j(k) \in N(i)$, i.e. the directed graph $\mathcal{G}_k$ mentioned above is equal to the interaction graph of $M(k)$.

Note that continuity of f_i is not needed in writing the state equations as Eq. (6.6). Another benefit of writing the system as Eq. (6.6) is that a lower bound on the convergence rate can be derived using Hajnal's inequality (Lemma 6.15).

7.4 Agreement in random networks

Consider the system in Eq. (6.6) where the stochastic matrices $M(k)$ are chosen according to some random process. This scenario was first studied in [Hatano and Mesbahi (2004)] using techniques in stochastic control. We derive sufficient conditions for agreement that depend on the types of graphs that have nonvanishing probabilities. For instance, if a scrambling graph occurs with nonzero probability, then the system synchronizes.

To formalize the problem, we consider the set of n by n stochastic matrices along with a probability measure on this set. We study synchronization of Eq. (6.6) when each $M(k)$ is taken independently from this set using the corresponding probability measure.

Definition 7.13 Eq. (6.6) synchronizes in probability if for any $x(0)$ and any $\epsilon > 0$,

$$Pr(\kappa(x(k)) \geq \epsilon) \to 0$$

as $k \to \infty$.

Theorem 7.14 *If there exists a compact set of stochastic scrambling matrices H such that $Pr(H) > 0$, then Eq. (6.6) synchronizes in probability.*

Proof: Pick $\epsilon, \psi > 0$. Since $Pr(H) > 0$, for any N there exists K such that for all $k \geq K$, at least N matrices in the set $\{M(2), \ldots, M(k)\}$ belongs to H with probability at least $1 - \psi$. Let $B = M(k)M(k-1)\cdots M(1)$. Let $\chi = \inf_{X \in H} \chi(X)$. By Lemma 6.15, $\delta(B) \leq (1-\chi)^N \delta(M(1))$. Since H is compact, $\chi > 0$ and this means that $\delta(B)$ can be made arbitrarily small for large enough N. Since $x(k+1) = (B \otimes \Pi_{i=1}^k D(i))x(1) + \mathbf{1} \otimes v'(k)$ for some vector $v'(k)$, $\|x_i(k+1) - x_j(k+1)\| = \| \sum_l (B_{il} - B_{jl}) \Pi_{i=1}^k D(i) x_l(1)\| \leq \delta(B) \sum_l \|x_l(1)\|$. This implies that with probability $1 - \psi$, $\kappa(x(k+1)) \leq \epsilon$ for large enough N. □

Suppose now that to each unweighted graph $\mathcal{G}$, there corresponds a stochastic matrix M such that the graph of M ignoring the weights is $\mathcal{G}$. We then consider Eq. (6.6) where at each k, a graph $\mathcal{G}$ is chosen independently from a random graph process and $M(k)$ is set to be the stochastic matrix corresponding to $\mathcal{G}$.

Corollary 7.15 *If the random graph model is $G_2(n, p)$ with $p > 0$, then Eq. (6.6) synchronizes in probability.*

Proof: Follows from Theorem 7.14 and the fact that the complete graph, which is scrambling, has nonzero probability in $G_2(n, p)$. □

Corollary 7.15 was shown in [Hatano and Mesbahi (2004)] for the case $D(k) = 1$ using stochastic Lyapunov theory.

Corollary 7.16 *If the random graph model is $G_1(n, M)$ with $M \geq 2n-3$, then Eq. (6.6) synchronizes in probability.*

Proof: Consider the simple graph $\mathcal{G}$ with $2n-3$ edges as shown in Figure 7.2. It is clear that every pair of vertices has a common child, and therefore this graph is scrambling. Since any graph with $\mathcal{G}$ as a subgraph is scrambling, the result follows from Theorem 7.14. □

Similar arguments show that these results hold for the digraph case as well.

Corollary 7.17 *If the random graph model is $G_{d2}(n, p)$ with $p > 0$, then Eq. (6.6) synchronizes in probability.*

Corollary 7.18 *If the random graph model is $G_{d1}(n, M)$ with $M \geq 2n - 1$, then Eq. (6.6) synchronizes in probability.*

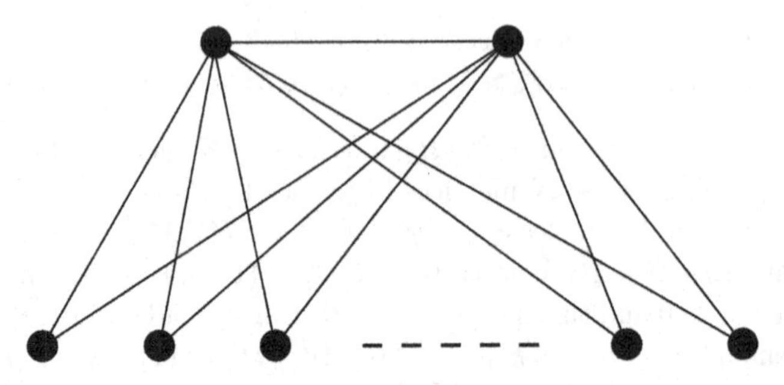

Fig. 7.2 A simple undirected scrambling graph.

Proof: The proof is the same as Corollary 7.16 except that we consider the simple scrambling directed graph with $2n - 1$ edges in Figure 7.3. □

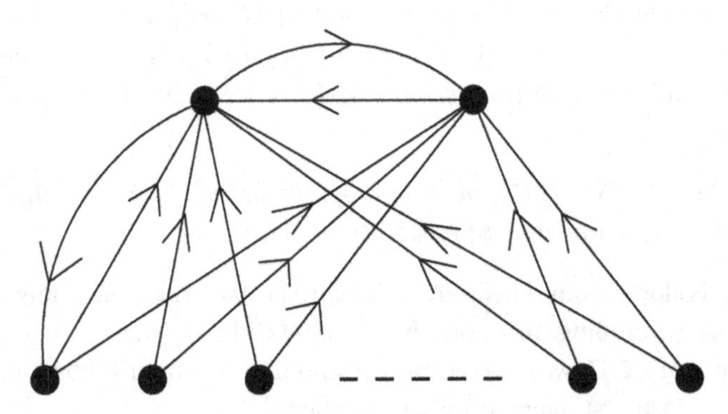

Fig. 7.3 A simple directed scrambling graph.

7.4.1 *Synchronization in random networks without the scrambling condition*

In this section, we study a synchronization condition that does not require the existence of a scrambling matrix with nonvanishing probability by restricting the matrices to have positive diagonal elements.

Theorem 7.19 *Consider n_p compact sets of matrices $S_i \subset S_d$ with $Pr(S_i) > 0$ for each $i = 1, \ldots, n_p$. Suppose that if $M_i \in S_i$, then the the union of the interaction graphs M_i, $i = 1, \ldots, n_p$, contains a spanning directed tree. Then Eq. (6.6) synchronizes in probability.*

Proof: The proof is similar to Theorem 7.14. In this case we choose K large enough such that a sequence of matrices

$M_{1,1},\ldots,M_{1,n_p},M_{2,1},\ldots,M_{2,n_p},\ldots,M_{n-1,1},\ldots M_{n-1,n_p}$ where $M_{i,j}\in S_j$ can be found as a subsequence of $M(2),\ldots,M(K)$ with probability at least $1-\psi$. Since the interaction graph of $M_{i,1}M_{i,2}\ldots M_{i,n_p}$ contains a spanning directed tree, by Lemma 6.28 and Corollary 6.30 $\Pi_{i,j}M_{i,j}$ is scrambling. Since the sets S_i are compact, $\chi(\Pi_{i,j}M_{i,j})$ is bounded away from 0. By choosing K even larger, we can make this happen at least N times. The rest of the proof is the same as that of Theorem 7.14. □

Corollary 7.20 *Suppose that to each simple graph $\mathcal{G}$, there corresponds a matrix $M\in S_d$ such that the graph of G minus the diagonal elements and ignoring the weights is equal to $\mathcal{G}$. If the random graph model is $G_1(n,K)$ or $G_{d1}(n,K)$ with $K>0$, then Eq. (7.2) where the matrices $M(k)$ are chosen according to this correspondence synchronizes in probability.*

Proof: Follows from Theorem 7.19 by choosing each set S_i to consist of the matrix corresponding to a single graph of M edges. □

Theorems 7.14 and 7.19 are applicable to other random graph models besides the classical models in Section 3.2. For instance, it is applicable to the small world and scale free random graph models in Chapter 3.

7.5 Further reading

Some early work in this area can be found in [Jadbabaie *et al.* (2003); Olfati-Saber and Murray (2004)]. Theorem 7.9 is a generalization of results in [Jadbabaie *et al.* (2003); Moreau (2003)]. This is a very active research area, see for instance [Ren *et al.* (2004); Lin *et al.* (2005); Ren and Beard (2005); Cao *et al.* (2005)]. See [Wu (2005a)] for a discussion of the connection between consensus problems and synchronization.

Appendix A

Algebraic Connectivity and Combinatorial Properties of a Graph

We summarize some of the properties of the various notions of algebraic connectivity and their relationships with combinatorial properties of the graph.

A.1 Properties of algebraic connectivity

Definition A.1 Let A be the adjacency matrix of a graph and S_1 and S_2 be subsets of vertices. $e(S_1, S_2)$ is defined as

$$e(S_1, S_2) = \sum_{v_1 \in S_1, v_2 \in S_2} A_{v_1, v_2}$$

which is the sum of the weights of edges which start in S_1 and ends in S_2.

In general, $e(S_1, S_2) \neq e(S_2, S_1)$ unless the graph is balanced. Since $A_{ij} \leq 1$, $e(S_1, S_2) \leq |S_1||S_2|$. The following result is quite useful and will be used throughout.

Lemma A.2 *Let S_1 and S_2 be two nontrivial disjoint subsets of vertices of a graph $\mathcal{G}$ (i.e. $0 < |S_1|$, $0 < |S_2|$ and $S_1 \cap S_2 = \emptyset$) and $\overline{S_i} = V \backslash S_i$. Then*

$$\begin{aligned} a_1(\mathcal{G}) &\leq \frac{|S_2|^2 e(S_1, \overline{S}_1) + |S_1||S_2|(e(S_1, S_2) + e(S_2, S_1)) + |S_1|^2 e(S_2, \overline{S}_2)}{|S_1||S_2|^2 + |S_1|^2|S_2|} \\ &\leq b_1(\mathcal{G}) \end{aligned}$$

If all weights are nonnegative, then

$$a_1(\mathcal{G}) \leq \frac{e(S_1, \overline{S}_1)}{|S_1|} + \frac{e(S_2, \overline{S}_2)}{|S_2|}$$

$$b_1(\mathcal{G}) \geq \frac{e(S_1,S_2)}{|S_1|} + \frac{e(S_2,S_1)}{|S_2|}$$

Proof: Let x be a vector such that $x_v = |S_2|$ if $v \in S_1$, $x_v = -|S_1|$ if $v \in S_2$ and $x_v = 0$ otherwise. Then $x \perp e$ and $x^T x = |S_1||S_2|^2 + |S_1|^2|S_2|$.

$$x^T D x = |S_2|^2 \sum_{v\in S_1} d_o(v) + |S_1|^2 \sum_{v \in S_2} d_o(v)$$

$$x^T A x = |S_2|^2 e(S_1,S_1) - |S_1||S_2| e(S_1,S_2) - |S_1||S_2| e(S_2,S_1) + |S_1|^2 e(S_2,S_2)$$

Since $e(S,S) + e(S,\overline{S}) = e(S,V) = \sum_{v\in S} d_o(v)$, this implies that

$$\begin{aligned} x^T L x &= x^T D x - x^T A x \\ &= |S_2|^2 e(S_1,\overline{S}_1) + |S_1||S_2|(e(S_1,S_2) + e(S_2,S_1)) + |S_1|^2 e(S_2,\overline{S}_2) \end{aligned}$$

Since $a_1(\mathcal{G}) \leq \frac{x^T L x}{x^T x} \leq b_1(\mathcal{G})$, the first set of inequalities follows. If all weights are nonnegative, the last 2 inequalities follow from the fact that $e(S_1,\overline{S}_1) \geq e(S_1,S_2)$ and $e(S_2,\overline{S}_2) \geq e(S_2,S_1)$. □

Corollary A.3 *Let S be a nontrivial subset of vertices of a graph $\mathcal{G}$ (i.e. $0 < |S| < n$)and $\overline{S} = V\backslash S$. Then*

$$a_1(\mathcal{G}) \leq \frac{e(S,\overline{S})}{|S|} + \frac{e(\overline{S},S)}{n-|S|} \leq b_1(\mathcal{G})$$

$$a_1(\mathcal{G}) \leq \frac{e(S,\overline{S})}{|S|} + |S|$$

Proof: Follows from Lemma A.2 and choosing $S_1 = S$, $S_2 = \overline{S}$. The last inequality follows from $e(S,\overline{S}) \leq |S||\overline{S}|$. □

Theorem A.4

(1) a_1 can be efficiently computed as

$$a_1(L) = \min_{x\in\mathbb{R}^{n-1}, \|Qx\|=1} x^T Q^T L Q x = \lambda_{\min}\left(\frac{1}{2} Q^T \left(L + L^T\right) Q\right)$$

where Q is an n by $n-1$ matrix whose columns form an orthonormal basis of $\mathbf{1}^\perp$;

(2) *Let* $\mathbf{T} = \{x \in \mathbb{R}^V, x \notin span(\mathbf{1})\}$ *and let* $L_K = nI - \mathbf{1}\mathbf{1}^T$ *be the Laplacian matrix of the complete graph. If* $\mathcal{G}$ *is balanced,*

$$a_1(\mathcal{G}) = n \min_{x \in T} \frac{x^T L x}{x^T L_K x} \leq n \max_{x \in T} \frac{x^T L x}{x^T L_K x} = b_1(\mathcal{G});$$

(3) *If the reversal of the graph* $\mathcal{G}$ *does not contain a spanning directed tree, then* $a_1(\mathcal{G}) \leq 0$;

(4) *If the graph is not weakly connected, then* $a_1(\mathcal{G}) \leq 0$;

(5) *Let* $\mathcal{G}$ *be a balanced graph. Then* $a_1(\mathcal{G}) > 0 \Leftrightarrow \mathcal{G}$ *is connected* $\Leftrightarrow \mathcal{G}$ *is strongly connected;*

(6) **Super- and sub-additivity.** $a_1(\mathcal{G} + \mathcal{H}) \geq a_1(\mathcal{G}) + a_1(\mathcal{H})$ *and* $b_1(\mathcal{G} + \mathcal{H}) \leq b_1(\mathcal{G}) + b_1(\mathcal{H})$;

(7) $a_1(\mathcal{G} \times \mathcal{H}) \leq \min(a_1(\mathcal{G}), a_1(\mathcal{H})) \leq \max(b_1(\mathcal{G}), b_1(\mathcal{H})) \leq b_1(\mathcal{G} \times \mathcal{H})$;

(8) $\lambda_1\left(\frac{1}{2}(L + L^T)\right) \leq a_1(L) \leq \lambda_2\left(\frac{1}{2}(L + L^T)\right)$, $\lambda_{n-1}\left(\frac{1}{2}(L + L^T)\right) \leq b_1(L) \leq \lambda_n\left(\frac{1}{2}(L + L^T)\right)$;

(9) *If* $\overline{L}$ *is the Laplacian matrix of the complement* $\overline{\mathcal{G}}$, *then* $a_1(L) + a_1(\overline{L}) = n$, *where* n *is the order of* L;

(10) *If the off-diagonal elements of the adjacency matrix* A *of* $\mathcal{G}$ *are random variables chosen independently according to* $P(A_{ij} = 1) = p$, $P(A_{ij} = 0) = 1 - p$, *then* $a_1(\mathcal{G}) \approx pn$ *in probability as* $n \to \infty$.

Proof: (1): Follows from the facts that $x \in \mathbf{K}$ if and only if $Qy = x$ for some $y \in \mathbb{R}^{n-1}$ and that $\|Qx\| = 1$ if and only if $\|x\| = 1$.
(2): Decompose $x \in \mathbf{T}$ as $x = \alpha\mathbf{1} + y$, where $y \perp \mathbf{1}$. Since $\mathbf{1}^T L = L\mathbf{1} = \mathbf{1}^T L_K = L_K \mathbf{1} = 0$, the conclusion follows from

$$\frac{x^T L x}{x^T L_K x} = \frac{y^T L y}{y^T L_K y} = \frac{y^T L y}{n y^T y}$$

(3-4): By Theorem 2.1 there exists a spanning directed tree in the reversal of $\mathcal{G}$ if and only if for any pair of vertices v and w, there exists a vertex z such that there is a directed path from v to z and a directed path from w to z. If the reversal of $\mathcal{G}$ does not have a spanning directed tree, then there exist a pair of vertices v and w such that for all vertices z, there is either no directed paths from v to z or no directed paths from w to z. Let $R(v)$ and $R(w)$ be the set of vertices reachable from v and w respectively. Let $\mathcal{H}(v)$ and $\mathcal{H}(w)$ be the subgraphs of $\mathcal{G}$ corresponding to $R(v)$ and $R(w)$ respectively. Express the Laplacian matrix of $\mathcal{H}(v)$ in Frobenius normal form (Eq. (2.1)). Let $B(v) = B_k$ be the square irreducible matrix in the lower right corner. We define $B(w)$ similarly. Note that $B(w)$ and $B(v)$

are zero row sums singular matrices. By the construction, it is easy to see that $B(v) = B_i$ and $B(w) = B_j$ in the Frobenius normal form (Eq. (2.1)) of the Laplacian matrix of G for some i, j. This means that $B_{ik} = 0$ for $k > i$ and $B_{jk} = 0$ for $k > j$.

To B_i and B_j correspond two nontrivial disjoint subsets of vertices S_1, S_2 of $\mathcal{G}$ such that the edges starting in S_i do not point outside of S_i, i.e. $A_{vw} \neq 0$ and $v \in S_i \Rightarrow w \in S_i$. The proof then follows from Lemma A.2 as $e(S_i, \overline{S}_i) = 0$ and $e(S_i, S_j) = 0$.

(5): If $\mathcal{G}$ is connected, then $L + L^T$ is irreducible. By Theorem 2.7 and Lemma 2.34, $a_1(\mathcal{G}) = \frac{1}{2}\lambda_2(L + L^T) > 0$. This together with (4) implies $a_1(\mathcal{G}) > 0 \Leftrightarrow G$ is connected. If $\mathcal{G}$ is balanced, then $\mathbf{1}^T L = 0$ and Lemma 2.28 shows that connectedness implies strongly connectedness.

(6): Since $L(\mathcal{G} + \mathcal{H}) = L(\mathcal{G}) + L(\mathcal{H})$,

$$\begin{aligned} a_1(\mathcal{G} + \mathcal{H}) &= \min_{x \in \mathbf{K}} x^T L(\mathcal{G} + \mathcal{H}) x \geq \min_{x \in \mathbf{K}} x^T L(\mathcal{G}) x + \min_{x \in \mathbf{K}} x^T L(\mathcal{H}) x \\ &= a_1(\mathcal{G}) + a_1(\mathcal{H}) \end{aligned}$$

The proofs of the other inequalities are similar.

(7): The Laplacian matrix L of $\mathcal{G} \times \mathcal{H}$ is $L(\mathcal{G}) \otimes I + I \otimes L(\mathcal{H})$ [Alon and Milman (1985)]. Let $x^T L(\mathcal{G}) x = a_1(\mathcal{G})$ and $y^T L(\mathcal{H}) y = a_1(\mathcal{H})$, $x \in P, y \in P$. Since $\frac{1}{\sqrt{n}} x \otimes \mathbf{1} \in P$, $a_1(\mathcal{G} \times \mathcal{H}) \leq (\frac{1}{\sqrt{n}} x \otimes \mathbf{1})^T L(\frac{1}{\sqrt{n}} x \otimes \mathbf{1}) = \frac{1}{n} x^T L(\mathcal{G}) x \otimes \mathbf{1}^T \mathbf{1} = a_1(\mathcal{G})$. Similarly using the vector $\frac{1}{\sqrt{n}} \mathbf{1} \otimes y$, we get $a_1(\mathcal{G} \times \mathcal{H}) \leq a_1(\mathcal{H})$. The proof for $b_1(\mathcal{G} \times \mathcal{H})$ is similar.

(8) Follows from the Courant-Fischer minmax theorem.

(9) Follows from the facts that $L(\mathcal{G}) + L(\overline{\mathcal{G}}) = L_K$ and $x^T L_K x = n$ for all $x \in \mathbf{K}$.

(10) Consider the symmetric matrix $B = \frac{1}{2}(A + A^T)$. Since $P(B_{ij} = 0) = q^2$, $P(B_{ij} = \frac{1}{2}) = 2pq$, $P(B_{ij} = 1) = p^2$, by [Füredi and Komlós (1981)],

$$\max_{i \leq n-1} |\lambda_i(B)| = o(n^{\frac{1}{2}+\epsilon}) \text{ in probability.}$$

Let $C = \frac{1}{2}(L + L^T) - (D_B - B)$ where D_B is the diagonal matrix with the row sums of B on the diagonal. Note that e is an eigenvector of $D_B - B$ and thus $\min_{x \in P} x^T (D_B - B) x = \lambda_2(D_B - B)$. Consider the diagonal matrix $F = (D_B - p(n-1)I)$. As in [Juhász (1991)] the interlacing properties of eigenvalues of symmetric matrices implies that $|\lambda_2(D_B - B) - \lambda_2(p(n-1)I - B)| \leq \rho(F) \leq \|F\|_\infty$. An application of a generalization of Chernoff's inequality [Alon and Spencer (2000)] (also known as Hoeffding's inequality) shows that $P(\|F\|_\infty \geq K n^{\frac{1}{2}+\epsilon}) \leq \sum_i P(|F_{ii}| \geq K n^{\frac{1}{2}+\epsilon}) \leq n e^{-\beta n^{2\epsilon}}$ and thus $\|F\|_\infty = o(n^{\frac{1}{2}+\epsilon})$ in probability. Therefore $|\lambda_2(D_B - B) - p(n-1) + \lambda_{n-1}(B)| = o(n^{\frac{1}{2}+\epsilon})$, i.e. $|\lambda_2(D_B - B)| = pn + o(n^{\frac{1}{2}+\epsilon})$. Next note that $C = D - D_B$ is a diagonal matrix and

$$a_1(\mathcal{G}_d) = \min_{x \in P} x^T L x \leq \lambda_2(D_B - B) + \max_{x \in P} x^T C x \leq \lambda_2(D_B - B) + \max_i C_{ii}$$

Similarly,

$$a_1(\mathcal{G}_d) \geq \lambda_2(D_B - B) + \min_{x \in P} x^T C x \geq \lambda_2(D_B - B) + \min_i C_{ii}$$

i.e. $|a_1(\mathcal{G}_d) - \lambda_2(D_B - B)| \leq \|C\|_\infty$. Similar applications of Hoeffding's inequality show that $\|C\|_\infty = o(n^{\frac{1}{2}+\epsilon})$ in probability which implies that $\|F\|_\infty + \|C\|_\infty = o(n^{\frac{1}{2}+\epsilon})$ in probability and thus the theorem is proved. □

Theorem A.5 *Consider a graph $\mathcal{G}$ with adjacency matrix A and Laplacian matrix L.*

(1) Let v,w be two vertices which are not adjacent, i.e. $A_{vw} = A_{wv} = 0$. Then

$$a_1(L) \leq \frac{1}{2}(d_o(v) + d_o(w)) \leq b_1(L)$$

In particular, if the graph has two vertices with zero outdegrees, then $a(L) \leq 0$.

(2)

$$\begin{aligned} a_1(L) &\leq \min_{v \in V} \left\{ d_o(v) + \frac{1}{n-1} d_i(v) \right\} \\ &\leq \max_{v \in V} \left\{ d_o(v) + \frac{1}{n-1} d_i(v) \right\} \leq b_1(L) \end{aligned}$$

(3)

$$a_1(L) \leq \min \left\{ \delta_o + \frac{1}{n-1}\Delta_i, \Delta_o + \frac{1}{n-1}\delta_i \right\} \leq \frac{n}{n-1} \min\{\Delta_o, \Delta_i\}$$

(4)

$$b_1(L) \geq \max \left\{ \delta_o + \frac{1}{n-1}\Delta_i, \Delta_o + \frac{1}{n-1}\delta_i \right\} \geq \frac{n}{n-1} \max\{\delta_o, \delta_i\}$$

(5)

$$\frac{1}{2}\min_{v\in V}\{d_o(v) - d_i(v)\} \le a_1(L) \le \min_{v\neq w}\{d_o(v) + d_o(w)\}$$

(6)

$$b_1(L) \le \max_{v\in V}\left\{\frac{3}{2}d_o(v) + \frac{1}{2}d_i(v)\right\}$$

(7) Let $\mathcal{H}$ be constructed from a graph $\mathcal{G}$ by removing a subset of vertices with zero indegree from $\mathcal{G}$ and all adjacent edges. Then $a_1(\mathcal{H}) \ge a_1(\mathcal{G})$.

(8) Let $\mathcal{H}$ be constructed from $\mathcal{G}$ by removing k vertices from $\mathcal{G}$ and all adjacent edges. Then $a_1(\mathcal{H}) \ge a_1(\mathcal{G}) - k$.

(9) For a graph $\mathcal{G}$, let (V_1, V_2) be a partition of V and let $\mathcal{G}_i$ be the subgraph generated from V_i. Then

$$a_1(\mathcal{G}) \le \min(a_1(\mathcal{G}_1) + |V_2|, a_1(\mathcal{G}_2) + |V_1|)$$

(10) If $\mathcal{G}$ is a directed tree and some vertex is the parent of at least two vertices, then $a_1(\mathcal{G}) \le 0$. If the reversal of $\mathcal{G}$ is a directed tree then $a_1(\mathcal{G}) \le \frac{d_i(r)}{n-1}$, where r is the root of the tree.

(11) A graph $\mathcal{G}$ with n vertices and m edges satisfies

$$a_1(\mathcal{G}) \le \left\lfloor \frac{m}{n} \right\rfloor + 1$$

Proof: (1) follows from Lemma A.2 and choosing $S_1 = \{v\}$, $S_2 = \{w\}$. (2) follows from Corollary A.3 and choosing $S = \{v\}$. We define $|E| = \sum_k d_i(k) = \sum_k d_o(k) \le n(n-1)$. Since

$$\begin{aligned}\min_{v\in V}\left\{d_o(v) + \tfrac{1}{n-1}d_i(v)\right\} &\le \min\{\delta_o + \tfrac{1}{n-1}\Delta_i, \Delta_o + \tfrac{1}{n-1}\delta_i\} \\ &\le \max\{\delta_o + \tfrac{1}{n-1}\Delta_i, \Delta_o + \tfrac{1}{n-1}\delta_i\} \\ &\le \max_{v\in V}\left\{d_o(v) + \tfrac{1}{n-1}d_i(v)\right\}\end{aligned}$$

and $\delta_o \le \frac{|E|}{n} \le \Delta_o$, $\delta_i \le \frac{|E|}{n} \le \Delta_i$, we have

$$a_1(\mathcal{G}) \le \frac{|E|}{n} + \frac{\Delta_i}{n-1} \le \frac{n}{n-1}\Delta_i$$

$$a_1(\mathcal{G}) \leq \frac{|E|}{n(n-1)} + \Delta_o \leq \frac{n}{n-1}\Delta_o$$

$$b_1(\mathcal{G}) \geq \frac{|E|}{n} + \frac{\delta_i}{n-1} \geq \frac{n}{n-1}\delta_i$$

$$b_1(\mathcal{G}) \geq \frac{|E|}{n(n-1)} + \delta_o \geq \frac{n}{n-1}\delta_o$$

which prove (3-4).
(5)-(6): The upper bound on $a_1(\mathcal{G})$ follows from Lemma A.2. Let $B = \frac{1}{2}(L + L^T)$. Then for each v, $B_{vv} + \sum_{v\neq w} |B_{vw}| = \frac{3}{2}d_o(v) + \frac{1}{2}d_i(v)$ and $B_{vv} - \sum_{v\neq w} |B_{vw}| = \frac{1}{2}d_o(v) - \frac{1}{2}d_i(v)$. By Theorem 2.9, $\lambda_{\max}(B) \leq \max_{v\in V}\left\{\frac{3}{2}d_o(v) + \frac{1}{2}d_i(v)\right\}$ and $\lambda_{\min}(B) \geq \min_{v\in V}\left\{\frac{1}{2}d_o(v) - \frac{1}{2}d_i(v)\right\}$. The result then follows from Theorem A.4(8).
(7): It is clear we only need to prove the case where one vertex of zero indegree is removed. Let $y \in \mathbf{K}$ be such that $a_1(\mathcal{H}) = y^T L(\mathcal{H})y$. The Laplacian matrix of $\mathcal{G}$ is of the form:

$$L(\mathcal{G}) = \begin{pmatrix} L(\mathcal{H}) & 0 \\ w^T & z \end{pmatrix}$$

Since $(y^T \quad 0)^T \in \mathbf{K}$ and $(y^T \quad 0)L(\mathcal{G})(y^T \quad 0)^T = a_1(\mathcal{H})$, it follows that $a_1(\mathcal{G}) \leq a_1(\mathcal{H})$.
(8): It is clear that we only need to consider the case $k = 1$. The Laplacian of graph $\mathcal{G}$ can be written as:

$$L(\mathcal{G}) = \begin{pmatrix} L(\mathcal{H}) + D & -v \\ w^T & z \end{pmatrix}$$

where D is a diagonal matrix with $D_{ii} = v_i \leq 1$. Define

$$F = \begin{pmatrix} L(\mathcal{H}) + I & -\mathbf{1} \\ w^T - (\mathbf{1}^T - v^T) & z + n - \sum_i v_i \end{pmatrix}$$

Let $y \in \mathbf{K}$ be such that $a_1(\mathcal{H}) = y^T L(\mathcal{H})y$. Then

$$\begin{aligned}(y^T \quad 0)F(y^T \quad 0)^T &= a_1(H) + 1 \geq \min_{x\in\mathbf{K}} x^T F x \\ &\geq \min_{x\in\mathbf{K}} x^T L(\mathcal{G})x + \min_{x\in\mathbf{K}} x^T (F - L(\mathcal{G}))x \\ &= a_1(G) + \min_{x\in\mathbf{K}} x^T (F - L(\mathcal{G}))x\end{aligned}$$

$$F - L(\mathcal{G}) = \begin{pmatrix} I - D & -(\mathbf{1} - v) \\ -(\mathbf{1}^T - v^T) & n - \sum_i v_i \end{pmatrix}$$

is a symmetric zero row sums matrix with nonpositive off-diagonal elements. Therefore $\min_{x \in \mathbf{K}} x^T(F - L(\mathcal{G}))x = \lambda_2(F - L(\mathcal{G})) \geq 0$ where the last inequality follows from Theorem 2.9.
(9) follows from (8).
(10): If $\mathcal{G}$ is a directed tree and some vertex is the parent of at least two vertices, consider the subtrees rooted at these two children with vertices S_1 and S_2. These two sets of vertices satisfy the condition in the proof of Theorem A.4(3) and thus $a_1(\mathcal{G}) \leq 0$. If the reversal of $\mathcal{G}$ is a directed tree, then (2) applied to the root results in the upper bound $\frac{d_i(r)}{n-1}$.
(11): let v be the vertex of graph $\mathcal{G}$ with minimum in-degree δ_i. This means that δ_i is less than or equal to the average in-degree of $\mathcal{G}$, i.e. $\delta_i \leq \lfloor \frac{m}{n} \rfloor$. Let $S = \{v\}$. Then $|S| = 1$

$$e(\overline{S}, S) = d_i(v) \leq \left\lfloor \frac{m}{n} \right\rfloor$$

and the result follows from Corollary A.3. □

Theorem A.6

(1) a_2 *can be efficiently computed as*

$$a_2(L) = \min_{x \in \mathbb{R}^{n-1}, \|Qx\|=1} x^T Q^T W L Q x = \lambda_{\min}\left(\frac{1}{2} Q^T \left(WL + L^T W\right) Q\right)$$

where Q is an n by $n-1$ matrix whose columns form an orthonormal basis of $\mathbf{1}^\perp$;

(2) If the reversal of the graph does not contain a spanning connected tree, then $a_2(L) \leq 0$;

(3) If the graph is strongly connected, then $0 < a_2 \leq Re(\lambda)$ for all eigenvalues λ of L not belonging to the eigenvector $\mathbf{1}$;

(4) If $\mathcal{G}$ is strongly connected, then

$$a_2(\mathcal{G}) \geq \frac{1 - \cos(\frac{\pi}{n})}{r} e(\mathcal{G})$$

$$a_2(\mathcal{G}) \geq \frac{C_1 e(\mathcal{G})}{2r} - C_2 q$$

where $r = \frac{\max_v w(v)}{\min_v w(v)}$, $q = \max_v w(v) d_o(v)$, $C_1 = 2(\cos(\frac{\pi}{n}) - \cos(\frac{2\pi}{n}))$, $C_2 = 2\cos(\frac{\pi}{n})(1 - \cos(\frac{\pi}{n}))$ and $e(\mathcal{G})$ is the edge connectivity (Definition A.12).

Proof: (1) follows from Theorem 2.35 since $a_2(L) = a_1(WL)$. (2): By Theorem 2.1, if the reversal of the graph does not contain a spanning directed tree, then there exist a pair of vertices v and w such that for all vertices z, either there are no directed paths from v to z or there are no directed paths from w to z. Let $R(v)$ and $R(w)$ denote the nonempty set of vertices reachable from v and w respectively. The hypothesis implies that $R(v)$ and $R(w)$ are disjoint. Furthermore, by definition $e(R(v), \overline{R(v)}) = e(R(w), \overline{R(w)}) = 0$ and the result follows from Lemma A.2 by setting $S_1 = R(v)$ and $S_2 = R(w)$. (3): if the graph is strongly connected, then $WL + L^TW$ is irreducible. Therefore by Theorem 2.7, $a_2 = \lambda_2(\frac{1}{2}(WL + L^TW)) > 0$. The proof of (4) is similar to [Fiedler (1973)]. Let $K = I - \frac{1}{2q}(WL + L^TW)$ where $q = \max_v\{w(v)d_o(v)\}$. K is doubly stochastic and by [Fiedler (1972)]

$$\frac{a_2(\mathcal{G})}{q} = 1 - \lambda_2(K) \geq 2\left(1 - \cos\left(\frac{\pi}{n}\right)\right)\mu$$

$$1 - \lambda_2(K) \geq C_1\mu - C_2$$

where $\mu = \min_{0<|S|<n}\{\sum_{v\in S, w\in \overline{S}} K_{vw}\}$. Note that

$$\mu q = \frac{1}{2}\min_{0<|S|<n}\left\{\sum_{v\in S} w(v)e(v,\overline{S}) + \sum_{v\in\overline{S}} w(v)e(v,S)\right\} \geq \frac{1}{2r}e(G)$$

and the result follows. □

Definition A.7 For an irreducible square matrix B with nonpositive off-diagonal elements, the functions $\beta(B)$ and $\gamma(B)$ are defined as follows. Decompose B uniquely as $B = L+D$, where L is a zero row sum matrix and D is a diagonal matrix. Let w be the unique positive vector such that $w^TL = 0$ and $\max_v w_v = 1$. The vector w exists by Perron-Frobenius theory (Theorem 2.7). Let $W = \mathrm{diag}(w)$. Then $\gamma(B) = \min_{x\neq 0, x\perp \mathbf{1}} \frac{x^TWBx}{x^T\left(W - \frac{ww^T}{\sum_v w_v}\right)x}$ and $\beta(B) = \min_{x\neq 0}\frac{x^TWBx}{x^TWx}$.

Lemma A.8 *Let B be an irreducible matrix with nonpositive off-diagonal elements and nonnegative row sums with decomposition $B = L + D$ as in Definition A.7. Then $\gamma(B) > 0$, $\beta(B) \geq 0$. Furthermore, $\beta(B) > 0$ if and only if $D \neq 0$.*

Proof: It is easy to see that $D \succeq 0$, $\gamma(B) \geq \min_{x \perp \mathbf{1}, x \neq 0} \frac{x^T WBx}{x^T x} = \frac{1}{2}\lambda_2(WB + B^T W)$ and $\beta(B) \geq \min_{x \neq 0} \frac{x^T WBx}{x^T x} = \frac{1}{2}\lambda_{\min}(WB + B^T W)$ where $\lambda_{\min}$ and λ_2 denote the smallest and the second smallest eigenvalue respectively. Since WL has zero column sums, $WL + L^T W$ is a zero row sums matrix. As w is a positive vector, this in turns implies that $WB + B^T W$ is irreducible and have nonnegative row sums and thus $\gamma(B) \geq \frac{1}{2}\lambda_2(WB + B^T W) > 0$ (Theorem 2.7). By Gershgorin circle criterion (Theorem 2.9), $\beta(B) \geq \frac{1}{2}\lambda_{\min}(WB + B^T W) \geq 0$. Suppose that $D = 0$. Then WB is a zero row sums and zero column sums matrix so that $\mathbf{1}^T WB = WB\mathbf{1} = 0$ and thus $\beta(B) = 0$. If $D \neq 0$, there exists i such that $D_{ii} > 0$ and thus $(WB + B^T W)_{ii} > |\sum_{j \neq i}(WB + B^T W)_{ij}|$. By Theorem 2.10, $WB + B^T W$ is nonsingular and thus $\beta(B) > 0$. □

Theorem A.9

(1) If the graph is strongly connected, then $a_3 \geq a_2 > 0$*;*
(2) $a_4 > 0$ *if and only if the reversal of the graph contains a spanning directed tree.*

Proof: (1) First note that $U = W - \frac{ww^T}{\|w\|_1}$ is a symmetric zero row sums matrix with nonpositive off-diagonal elements and is thus a positive semidefinite matrix. Since $x^T Ux \leq x^T Wx \leq x^T x$, this implies that $a_3 \geq a_2$. (2) Let V_i be the subset of vertices corresponding to B_i. If the reversal of the graph contains a spanning directed tree, then its root must be in V_k. Furthermore, there is a directed path from every other vertex to the root. If $D_i = 0$ for some $i < k$, then there are no paths from V_i to V_k, a contradiction. Therefore by Lemma A.8 $\eta(L) > 0$.

If the reversal of the graph does not have a spanning directed tree, then by Theorem 2.1 there exist a pair of vertices v and w such that for all vertices z, there is either no directed paths from v to z or no directed paths from w to z. Let $R(v)$ and $R(w)$ be the set of vertices reachable from v and w respectively, which must necessarily be disjoint. Let $H(v)$ and $H(w)$ be the subgraphs of G corresponding to $R(v)$ and $R(w)$ respectively. Expressing the Laplacian matrix of $H(v)$ in Frobenius normal form, let $B(v)$ be the square irreducible matrix in the lower right corner. We define $B(w)$ similarly. Note that $B(w)$ and $B(v)$ are zero row sums singular matrices. By the construction, it is easy to see that $B(v) = B_i$ and $B(w) = B_j$ in the Frobenius normal form (Eq. (2.1)) of L for some i, j. By Lemma A.8, $\beta(B_i) = \beta(B_j) = 0$ and thus $\eta(L) = 0$ since either $i \neq k$ or $j \neq k$. □

A.2 Vertex and edge connectivity

Definition A.10 The vertex connectivity $v(\mathcal{G})$ of a graph is defined as the size of the smallest subset of vertices such that its removal along with all adjacent edges results in a disconnected graph. If no such vertex set exists, $v(\mathcal{G}) = n$.

Theorem A.11 $a_1(\mathcal{G}) \leq v(\mathcal{G})$.

Proof: This is a direct consequence of Theorem 2.36(8) and Theorem 2.35(4). □

Definition A.12 The edge connectivity $e(\mathcal{G})$ of a graph is defined as the smallest weighted sum among all subsets of edges such that its removal results in a disconnected graph.

Theorem A.13 *For a graph with nonnegative weights,* $a_1(\mathcal{G}) \leq e(\mathcal{G})$.

Proof: Let S be a connected component of the disconnected graph resulting from removal of a minimal set of edges. Then $e(S,\overline{S}) + e(\overline{S},S) = e(\mathcal{G})$. By Corollary A.3, $a_1(\mathcal{G}) \leq \frac{e(S,\overline{S})}{|S|} + \frac{e(\overline{S},S)}{|\overline{S}|} \leq e(S,\overline{S}) + e(\overline{S},S)$. □

A.3 Graph partitions

A.3.1 *Maximum directed cut*

Definition A.14 The maximum directed cut $md(\mathcal{G})$ is defined as:

$$md(\mathcal{G}) = \max_{0<|S|<n} \{e(S,\overline{S})\}$$

Theorem A.15 *For a graph with nonnegative weights,*

$$md(\mathcal{G}) \leq (n-1)\min(b_1(\mathcal{G}), b_1(\mathcal{G}^R))$$

$$md(\mathcal{G}) \leq (n-2)b_1(\mathcal{G}) + \max(0, b(\mathcal{G}) - (n-1)\delta_o)$$

where $\mathcal{G}^R$ *is the reversal of* $\mathcal{G}$.

Proof: Since $|S| \leq n-1$, by Corollary A.3 we have

$$\frac{e(S,\overline{S})}{n-1} \leq \frac{e(S,\overline{S}) + e(\overline{S},S)}{n-1} \leq \frac{e(S,\overline{S})}{|S|} + \frac{e(\overline{S},S)}{|\overline{S}|} \leq b_1(\mathcal{G})$$

Similarly, $\frac{e(S,\overline{S})}{n-1} \leq b_1(\mathcal{G}^R)$. If $md(\mathcal{G})$ is achieved with $|S| \leq n-2$, then $md(\mathcal{G}) \leq (n-2)b_1(G)$. If $md(\mathcal{G})$ is achieved with $|S| \leq n-1$, then $b_1(G) \geq \frac{e(S,\overline{S})}{n-1} + \delta_o$, so in either case $md(\mathcal{G}) \leq (n-2)b_1(\mathcal{G}) + \max(0, b_1(\mathcal{G}) - (n-1)\delta_o)$. □

Note that for the imploding star graph (Fig. 2.6) the bound in Theorem A.15 is tight.

A.3.2 *Edge-forwarding index*

The definition of edge-forwarding index in [Heydemann *et al.* (1989)] can also be applied to directed graphs.

Definition A.16 Given a strongly connected unweighted directed graph, a routing is defined as a set of $n(n-1)$ paths $R(u,v)$ between any pair of distinct vertices v, w of $\mathcal{G}$. The load of an edge e, $\pi(\mathcal{G}, R, e)$, is defined as the number of paths in the routing R which traverse it. The edge-forwarding index of $(\mathcal{G}, R)$ is defined as $\pi(\mathcal{G}, R) = \max_{e\in E}(\mathcal{G}, R, e)$. The edge-forwarding index of the graph $\mathcal{G}$ is defined as $\pi(\mathcal{G}) = \min_R \pi(\mathcal{G}, R)$.

Theorem A.17 *Let $\mathcal{G}$ be a strongly connected unweighted directed graph. For $S \subset V$,*

$$\pi(\mathcal{G}) \geq \max\left(\frac{|S|(n-|S|)}{e(S,\overline{S})}, \frac{|S|(n-|S|)}{e(\overline{S},S)}\right) \geq \frac{n}{b_1(\mathcal{G})}$$

Proof: The proof is similar to [Mohar (1997)]. Let R be a routing. Each path in R from vertex v in S to vertex w in $\overline{S}$ contains at least one edge in the edge cut of S. Since there are $|S|(n-|S|)$ such paths, $\pi(\mathcal{G}) \geq \frac{|S|(n-|S|)}{e(S,\overline{S})}$. Similarly, $\pi(\mathcal{G}) \geq \frac{|S|(n-|S|)}{e(\overline{S},S)}$. Let $t = \min(e(S,\overline{S}), e(\overline{S},S))$. By Corollary A.3,

$$b_1(\mathcal{G}) \geq \frac{t}{|S|} + \frac{t}{n-|S|} = t\frac{n}{|S|(n-|S|)}$$

which implies the second inequality. □

A.3.3 *Bisection width*

Definition A.18 The bisection width is defined as:

$$bw(\mathcal{G}) = \min_{|S|=\lfloor \frac{n}{2} \rfloor} \{e(S,\overline{S})\}$$

A related quantity is

$$\overline{bw}(\mathcal{G}) = \max_{|S|=\lfloor \frac{n}{2} \rfloor} \{e(S, \overline{S})\}$$

It is easy to see that

$$bw(\mathcal{G}) + \overline{bw}(\overline{\mathcal{G}}) = \left\lfloor \frac{n}{2} \right\rfloor \left\lceil \frac{n}{2} \right\rceil \tag{A.1}$$

For the exploding[3] and imploding star graphs with more than 2 vertices, $bw(\mathcal{G}) = 0$.

Theorem A.19

$$bw(\mathcal{G}) \geq \left\lfloor \frac{n}{2} \right\rfloor \left(a_1(\mathcal{G}) - \left\lfloor \frac{n}{2} \right\rfloor\right)$$

$$\overline{bw}(\mathcal{G}) \leq \left\lfloor \frac{n}{2} \right\rfloor b_1(\mathcal{G})$$

Proof: The first inequality follows from Corollary A.3 by setting $|S| = \lfloor \frac{n}{2} \rfloor$. The second inequality follows from Eq. (A.1) and Theorem A.4(9). □

As in Boppana (1987) (see also Mohar (1997)), the bound on the bisection width can be improved by the use of correction functions c.

Theorem A.20 *Let n be even. Then*

$$bw(\mathcal{G}) \geq \frac{n}{2}\left(a^*(\mathcal{G}) - \frac{n}{2}\right)$$

where

$$a^*(\mathcal{G}) = \max_{c \perp \mathbf{1}} \min_{x \in P} x^T(diag(c) + L(G))x$$

Proof: In the proof of Lemma A.2, for $|S_1| = |S_2| = \frac{n}{2}$, the vector x as defined satisfies $x_v^2 = x_w^2$ and thus $x^T \mathrm{diag}(c) x = \sum_{v \in V} c_v x_v^2 = 0$ implying $a^*(\mathcal{G}) \leq \frac{x^T L x}{x^T x}$. The rest of the proof is similar to that of Lemma A.2. □

A.3.4 *Isoperimetric number*

Definition A.21 The isoperimetric number $i(\mathcal{G})$ is defined as:

$$i(\mathcal{G}) = \min_{0 < |S| \leq \frac{n}{2}} \left\{ \frac{e(S, \overline{S})}{|S|} \right\}$$

[3] An exploding star graph is the reversal of an imploding star graph (Fig. 2.6).

Theorem A.22 *The isoperimetric number of a graph satisfies:*

$$i(\mathcal{G}) \geq a_1(\mathcal{G}) - \left\lfloor \frac{n}{2} \right\rfloor$$

Proof: Follows from Corollary A.3. □

A.3.5 *Minimum ratio cut*

Definition A.23 The minimum ratio cut $rc(\mathcal{G})$ is defined as [Hagen and Kahng (1992)]:

$$rc(\mathcal{G}) = \min_S \left\{ \frac{e(S, \overline{S})}{|S||\overline{S}|} \right\}$$

Theorem A.24 *The minimum ratio cut of a graph satisfies:*

$$rc(\mathcal{G}) \geq \frac{a_1(\mathcal{G}) - \lfloor \frac{n}{2} \rfloor}{\lfloor \frac{n}{2} \rfloor}$$

Proof: Follows from Corollary A.3. □

The lower bounds on the bisection width, isoperimetric number and minimum ratio cut in Theorems A.19-A.24 are nontrivial only when $a(\mathcal{G}) \geq \lfloor \frac{n}{2} \rfloor$, and they are tight for the union of $\lfloor \frac{n}{2} \rfloor$ imploding star graphs (Fig. 2.7). The Laplacian matrix of this graph can be written as $\lfloor \frac{n}{2} \rfloor I - [J \quad 0]$, where J is the n by $\lfloor \frac{n}{2} \rfloor$ matrix of all 1's. If $x \perp \mathbf{1}$, then $x^T J = 0$, and thus $x^T L x = \lfloor \frac{n}{2} \rfloor x^T x$. Therefore $a_1(\mathcal{G}) = b_1(\mathcal{G}) = \lfloor \frac{n}{2} \rfloor$. Since there are no edges out of V_1, $bw(\mathcal{G}) = i(\mathcal{G}) = rc(\mathcal{G}) = 0$.

A.3.6 *Independence number*

Definition A.25 An independent set of vertices is a set of vertices such that no two distinct vertices in the set are adjacent. The independence number of a graph $\alpha(\mathcal{G})$ is the size of the largest independent set of vertices.

Theorem A.26 *Let the indegrees and outdegrees be ordered as $d_o(1) \leq d_o(2) \leq \cdots \leq d_o(n)$, $d_i(1) \leq d_i(2) \leq \cdots \leq d_i(n)$ respectively and define $e_o(r) = \frac{1}{r}\sum_{j=1}^{r} d_o(j)$ and $e_i(r) = \frac{1}{r}\sum_{j=1}^{r} d_i(j)$. If r_0 is the smallest integer r such that*

$$r(b_1(G) + e_i(r) - e_o(r)) > n(b_1(\mathcal{G}) - e_o(r))$$

then $\alpha(\mathcal{G}) \leq r_0 - 1$.

Proof: Let S be an independent set such that $|S| = r$. Since S is independent, $e(S, \overline{S}) = \sum_{v \in S} d_o(v) \geq re_o(r)$ and $e(\overline{S}, S) = \sum_{v \in S} d_i(v) \geq re_i(r)$. Then by Corollary A.3

$$b_1(\mathcal{G}) \geq e_o(r) + \frac{re_i(r)}{n-r}$$

which implies that

$$r(b_1(\mathcal{G}) + e_i(r) - e_o(r)) \leq n(b_1(\mathcal{G}) - e_o(r))$$

□

For a graph $\mathcal{G}$, construct the unweighted undirected graph by ignoring the weight, multiplicity and orientation of each edge. It is clear that the independence number of these two graphs are the same. Thus Theorem A.26 suggests an algorithm for improving the upper bound of the independence number considered in [Mohar (1991b)]. Given an undirected unweighted graph $\mathcal{G}$ with adjacency matrix A, consider the class of graphs $\mathcal{U}$ whose edges are at the same place as $\mathcal{G}$, i.e. graphs in $\mathcal{U}$ has adjacency matrices $\tilde{A}$ such that $(\tilde{A}_{vw} \neq 0$ or $\tilde{A}_{wv} \neq 0) \Leftrightarrow A_{vw} \neq 0$. For a given r, find $\tilde{\mathcal{G}} \in \mathcal{U}$ such that $\frac{b_1(\tilde{\mathcal{G}}) - e_o(r)}{b_1(\tilde{\mathcal{G}}) + e_i(r) - e_o(r)}$ is minimized[4] and see if this reduces the value of r_0. If so, set r equal to r_0 and repeat. It is not clear what the best strategy is to find a good graph in $\mathcal{U}$ which reduces r_0.

Since vertex connectivity is also independent on the orientations and weights of the edges, similar statements can be made about the relationship between $a_1(\mathcal{G})$ and $v(\mathcal{G})$.

A.4 Semibalanced graphs

The bounds on the bisection width, maximum directed cut, the isoperimetric number and minimum ratio cut can be improved if the difference between $e(S, \overline{S})$ and $e(\overline{S}, S)$ is small, i.e. the graph is close to being balanced.

Definition A.27 A graph is (α, β)-semibalanced if $-\beta \leq d_i(v) - d_o(v) \leq \alpha$ for all $v \in V$.

Lemma A.28 *If $\mathcal{G}$ is (α, β)-semibalanced, then*

$$a_1(\mathcal{G}) \leq \frac{n}{n-1}\delta_o + \frac{\alpha}{n-1}$$

[4]where $e_o(r)$ and $e_i(r)$ are calculated using the in- and outdegrees of $\tilde{\mathcal{G}}$.

$$b_1(\mathcal{G}) \geq \frac{n}{n-1}\Delta_o - \frac{\beta}{n-1}$$

If in addition all weights are nonnegative,

$$a_1(\mathcal{G}) \geq -\frac{\alpha}{2}, \quad b_1(\mathcal{G}) \leq 2\Delta_o + \frac{\alpha}{2}$$

Proof: Follows from Theorem A.5. □

Since balanced graphs are $(0,0)$-semibalanced, Lemma A.28 is a generalization of Theorem 2.31(2).

Lemma A.29 *If $\mathcal{G}$ is (α,α)-semibalanced, then*

$$|e(S,\overline{S}) - e(\overline{S},S)| \leq \alpha \min(|S|, n-|S|)$$

Proof: Follows from the fact that $e(S,\overline{S}) - e(\overline{S},S) = \sum_{v\in S} d_o(v) - d_i(v)$. □

Theorem A.30 *If $\mathcal{G}$ is (α,α)-semibalanced with nonnegative weights, then*

$$md(\mathcal{G}) \leq \left\lfloor \frac{n}{2} \right\rfloor \left\lceil \frac{n}{2} \right\rceil \frac{b_1(\mathcal{G}) + \alpha}{n}$$

Proof: From Corollary A.3,

$$\begin{aligned} b_1(\mathcal{G}) &\geq \frac{e(S,\overline{S})}{|S|} + \frac{e(\overline{S},S)}{|\overline{S}|} \\ &= \left(\frac{1}{|S|} + \frac{1}{|\overline{S}|}\right) e(S,\overline{S}) + \frac{e(\overline{S},S) - e(S,\overline{S})}{|\overline{S}|} \\ &\geq \frac{n}{|S||\overline{S}|} e(S,\overline{S}) - \alpha \frac{\min(|S|,|\overline{S}|)}{|\overline{S}|} \geq \frac{n}{|S||\overline{S}|} e(S,\overline{S}) - \alpha \end{aligned}$$

Since

$$\frac{n}{|S|(n-|S|)} \geq \frac{n}{\left\lfloor \frac{n}{2} \right\rfloor \left\lceil \frac{n}{2} \right\rceil}$$

the result follows. □

Theorem A.31 *If $\mathcal{G}$ is (α,α)-semibalanced with nonnegative weights, then*

$$\overline{bw}(\mathcal{G}) \leq \left\lfloor \frac{n}{2} \right\rfloor \left\lceil \frac{n}{2} \right\rceil \frac{b_1(\mathcal{G}) + \alpha}{n}$$

$$bw(\mathcal{G}) \geq \left\lfloor \frac{n}{2} \right\rfloor \left\lceil \frac{n}{2} \right\rceil \frac{a_1(\mathcal{G}) - \alpha}{n}$$

Proof: The first inequality follows from a similar proof as Theorem A.30. The second inequality is due to Eq. (A.1) and Theorem A.4(9). □

Theorem A.32 *If G is (α, α)-semibalanced with nonnegative weights, then*

$$i(\mathcal{G}) \geq \frac{(a_1(\mathcal{G}) - \alpha)\left\lceil \frac{n}{2} \right\rceil}{n}$$

Proof: Similar to the proof of Theorem A.30,

$$a_1(\mathcal{G}) \leq \frac{n}{|S||\overline{S}|} e(S, \overline{S}) + \alpha$$

Since $|\overline{S}| \geq \left\lceil \frac{n}{2} \right\rceil$, the result follows. □

Theorem A.33 *If G is (α, α)-semibalanced with nonnegative weights, then*

$$rc(\mathcal{G}) \geq \frac{a_1(\mathcal{G}) - \alpha}{n}$$

Proof: The proof is similar to the proof of Theorem A.32. □

Bibliography

Agaev, R. and Chebotarev, P. (2005). On the spectra of nonsymmetric Laplacian matrices, *Linear Algebra and Its Applications* **399**, pp. 157–168.

Aiello, W., Chung, F. and Lu, L. (2001). A random graph model for power law graphs, *Experimental Mathematics* **10**, 1, pp. 53–66.

Alon, N. and Milman, V. D. (1985). λ_1, isoperimetric inequalities for graphs, and superconcentrators, *Journal of Combinatorial Theory, Series B* **38**, pp. 73–88.

Alon, N. and Spencer, J. (2000). *The Probabilistic Method*, 2nd ed. (Wiley).

Alsedà, L., Llibre, J., Misiurewicz, M. and Tresser, C. (1989). Periods and entropy for Lorenz-like maps, *Ann. Inst. Fourier (Grenoble)* **39**, 4, pp. 929–952.

Amano, M., Luo, Z.-W. and Hosoe, S. (2003). Graph dependant sufficient conditions for synchronization of dynamic network system with time-delay, in *Proceedings of 4th IFAC Workshop on Time Delay Systems (TDS '03)*.

Anthonisse, J. M. and Tijms, H. (1977). Exponential convergence of products of stochastic matrices, *Journal of Mathematical Analysis and Applications* **59**, pp. 360–364.

Artés, J. C., Grünbaum, B. and Llibre, J. (1997). On the invariant straight lines of polynomial differential systems, *Differential Equations Dynam. Systems* **5**, 3–4, pp. 317–327.

Atay, F. M., Biyikoğlu, T. and Jost, J. (2006). Synchronization of networks with prescribed degree distributions, *IEEE Transactions on Circuits and Systems-I: Fundamental Theory and Applications* **53**, 1, pp. 92–98.

Balmforth, N., Tresser, C., Worfolk, P. and Wu, C. W. (1997). Master-slave synchronization and the Lorenz equations, *Chaos* **7**, 3, pp. 392–394.

Banavar, J., Maritan, A. and Rinaldo, A. (1999). Size and form in efficient transportation networks, *Nature* **399**, pp. 130–132.

Bapat, R. B., Grossman, J. W. and Kulkarni, D. M. (1999). Generalized matrix tree theorem for mixed graphs, *Linear and Multilinear Algebra* **46**, pp. 299–312.

Barabási, A.-L. and Albert, R. (1999). Emergence of scaling in random networks, *Science* **286**, 5439, pp. 509–512.

Barabási, A.-L., Albert, R. and Jeong, H. (2000). Scale-free characteristics of random networks: the topology of the world wide web, *Physica A* **281**, pp. 69–77.

Barahona, M. and Pecora, L. M. (2002). Synchronization in small-world systems, *Physical Review Letters* **89**, 5, p. 054101.

Belykh, I. V., Belykh, V. N. and Hasler, M. (2004a). Blinking model and synchronization in small-world networks with a time-varying coupling, *Physica D* **195**, pp. 188–206.

Belykh, V. N., Belykh, I. V. and Hasler, M. (2004b). Connection graph stability method for synchronized coupled chaotic systems, *Physica D* **195**, pp. 159–187.

Bennett, M., Schatz, M. F., Rockwood, H. and Wiesenfeld, K. (2002). Huygens's clocks, *Proceedings of the Royal Society A* **458**, 2019, pp. 563–579.

Bolla, M. (1993). Spectra, euclidean representations and clusterings of hypergraphs, *Discrete Mathematics* **117**, pp. 19–39.

Bollobás, B. (1988). The isoperimetric number of random regular graphs, *European Journal of Combinatorics* **9**, pp. 241–244.

Bollobás, B. (2001). *Random Graphs*, 2nd ed. (Cambridge University Press).

Bollobás, B. and Riordan, O. M. (2004). Mathematical results on scale-free random graphs, in S. Bornholdt and H. G. Schuster (eds.), *Handbook of Graphs and Networks*, chap. 1 (Wiley-VCH), pp. 1–34.

Boppana, R. B. (1987). Eigenvalues and graph bisection: an average case analysis, in *28th IEEE Annual Symposium on the Foundations of Computer Science*, pp. 280–285.

Borchers, B. (1999). CSDP: a C library for semidefinite programming, *Optimization methods and Software* **11**, 1, pp. 613–623, `http://www.nmt.edu/~borchers/csdp.html`.

Braess, D. (1968). Über ein paradoxon aus der verkehrsplanung, *Unternehmensforschung* **12**, pp. 258–268, English translation in *Transportation Science*, vol. 39, pages 446-450, 2005.

Bru, R., Elsner, L. and Neumann, M. (1994). Convergence of infinite products of matrices and inner-outer iteration schemes, *Electronic Transactions on Numerical Analysis* **2**, pp. 183–193.

Brualdi, R. A. and Ryser, H. J. (1991). *Combinatorial Matrix Theory* (Cambridge University Press).

Cao, M., Morse, A. S. and Anderson, B. D. O. (2005). Coordination of an asynchronous multi-agent system via averaging, in *Proceedings of 2005 IFAC conference*.

Cao, M. and Wu, C. W. (2007). Topology design for fast convergence of network consensus algorithms, in *Proceedings of IEEE International Symposium on Circuits and Systems*.

Casselman, B. (2004). Networks, *Notices of the American Mathematical Society* **51**, 4, pp. 392–393.

Chaté, H. and Manneville, P. (1992). Collective behaviors in coupled map lattices with local and nonlocal connections, *Chaos* **2**, 3, pp. 307–313.

Chatterjee, S. and Seneta, E. (1977). Towards consensus: some convergence theorems on repeated averaging, *Journal of Applied Probability* **14**, pp. 89–97.

Chua, L. O. and Green, D. N. (1976). A qualitative analysis of the behavior of dynamic nonlinear networks: Stability of autonomous networks, *IEEE Transactions on Circuits and Systems* **23**, 6, pp. 355–379.

Chung, F. and Lu, L. (2002). Connected components in a random graph with given degree sequences, *Annals of combinatorics* **6**, pp. 125–145.

Chung, F. R. K., Faber, V. and Manteuffel, T. A. (1994). An upper bound on the diameter of a graph from eigenvalues associated with its Laplacian, *SIAM Journal on Discrete Mathematics* **7**, 3, pp. 443–457.

Chung, F. R. K., Graham, R. L. and Yau, S.-T. (1996). On sampling with Markov chains, *Random Structures and Algorithms* **9**, 1–2, pp. 55–77.

Coppersmith, D. and Wu, C. W. (2005). Conditions for weak ergodicity of inhomogeneous Markov chains, Research Report 23489, IBM.

Dall, J. and Christensen, M. (2002). Random geometric graphs, *Physical Review E* **66**, p. 016121.

de Solla Price, D. J. (1965). Networks of scientific papers, *Science* **149**, pp. 510–515.

de Solla Price, D. J. (1976). A general theory of bibliometric and other cumulative advantage processes, *J. Amer. Soc. Inform. Sci.* **27**, pp. 292–306.

DeGroot, M. H. (1974). Reaching a consensus, *Journal of the American Statistical Association* **69**, 345, pp. 118–121.

Desplanques, J. (1887). Théorème d'algébre, *J. de Math. Spec.* **9**, pp. 12–13.

Dmitriev, N. and Dynkin, E. (1945). On the characteristic numbers of stochastic matrices, *Dokl. Akad. Nauk USSR* **49**, pp. 159–162, in Russian.

Ellis, R. B., Martin, J. L. and Yan, C. (2006). Random geometric graph diameter in the unit ball, *Algorithmica* To appear.

Erdös, P. and Gallai, T. (1960). Graphs with prescribed degrees of vertices, *Mat. Lapok* **11**, pp. 264–274.

Erdös, P. and Graham, R. L. (1972). On sums of Fibonacci numbers, *Fibonacci Quarterly* **10**, 3, pp. 249–254.

Erdös, P. and Grünbaum, B. (1973). Osculation vertices in arrangements of curves, *Geometriae Dedicata* **1**, 3, pp. 322–333.

Erdös, P. and Renyi, A. (1959). On random graphs I, *Publicationes, Mathematicae* **6**, pp. 290–297.

Fang, L. and Antsaklis, P. (2005). On communication requirements for multi-agent consensus seeking, in *2005 Workshop on Networked Embedded Sensing and Control*.

Fax, J. A. (2002). *Optimal and Cooperative Control of Vehicle Formations*, Ph.D. thesis, California Institute of Technology.

Fiedler, M. (1972). Bounds for eigenvalues of doubly stochastic matrices, *Linear Algebra and Its Applications* **5**, pp. 299–310.

Fiedler, M. (1973). Algebraic connectivity of graphs, *Czechoslovak Mathematical Journal* **23**, 98, pp. 298–305.

Friedman, J., Kahn, J. and Szemerédi, E. (1989). On the second eigenvalue in random regular graphs, in *Proceedings of the 21st Annual ACM Symposium on Theory of Computing*, pp. 587–598.

Fujisaka, H. and Yamada, T. (1983). Stability theory of synchronized motion in coupled-oscillator systems, *Progress of Theoretical Physics* **69**, 1, pp. 32–47.

Füredi, Z. and Komlós, J. (1981). The eigenvalues of random symmetric matrices, *Combinatorica* **1**, pp. 233–241.

Gantmacher, F. R. (1960). *The Theory of Matrices*, vol. 1 (Chelsea Publishing Company, New York).

Geršgorin, S. A. (1931). Über die abgrenzung der eigenwerte einer matrix, *Izv. Akad. Nauk. USSR Otd. Fiz.-Mat. Nauk* **7**, pp. 749–754.

Gilbert, E. N. (1959). Random graphs, *The Annals of Mathematical Statistics* **30**, 4, pp. 1141–1144.

Gilbert, E. N. (1961). Random plane networks, *Journal of the Society for Industrial & Applied Mathematics* **9**, pp. 533–553.

Godsil, C. and Royle, G. (2001). *Algebraic Graph Theory*, Graduate Texts in Mathematics (Springer-Verlag).

Gradshteyn, I. S. and Ryzhik, I. M. (1994). *Table of Integrals, Series, and Products*, 5th ed. (Academic Press).

Greene, B. R., Shapere, A., Vafa, C. and Yau, S.-T. (1990). Stringy cosmic strings and noncompact Calabi-Yau manifolds, *Nuclear Physics B* **227**, 1, pp. 1–36.

Guattery, S., Leighton, F. T. and Miller, G. (1997). The path resistance method for bounding λ_2 of a Laplacian, in *Proceedings 6th ACM-SIAM Symposium on Discrete Algorithms*, pp. 201–210.

Guimerà, R., Mossa, S., Turtschi, A. and Amaral, L. A. N. (2005). The worldwide air transportation network: Anomalous centrality, community structure, and cities' global roles, *Proceedings National Academy of Sciences* **102**, 22, pp. 7794–7799.

Hagen, L. and Kahng, A. B. (1992). New spectral methods for ratio cut partitioning and clustering, *IEEE Transactions on Computer-Aided Design* **11**, 9, pp. 1074–1085.

Hajnal, J. (1958). Weak ergodicitiy in non-homogeneous Markov chains, *Proc. Cambridge Philos. Soc.* **54**, pp. 233–246.

Hakimi, S. L. (1962). On the realizability of a set of integers as degrees of the vertices of a graph, *SIAM J. Appl. Math.* **10**, pp. 496–506.

Hale, J. K. (1977). *Theory of Functional Differential Equations*, *Applied Mathematical Sciences*, vol. 3 (Springer-Verlag).

Hatano, Y. and Mesbahi, M. (2004). Agreement over random networks, *43rd IEEE Conference on Decision and Control* , pp. 2010–2015.

Havel, V. (1955). A remark on the existence of finite graphs, *Casopis Pest. Mat.* **80**, pp. 477–480.

Heydemann, M. C., Meyer, J. C. and Sotteau, D. (1989). On forwarding indices of networks, *Discrete Applied Mathematics* **23**, pp. 103–123.

Horn, R. A. and Johnson, C. R. (1985). *Matrix analysis* (Cambridge University Press).

Ilchmann, A., Owens, D. H. and Prätzel-Wolters, D. (1987). Sufficient conditions for stability of linear time-varying systems, *Systems and Control Letters* **9**, pp. 157–163.

Jadbabaie, A., Lin, J. and Morse, A. S. (2003). Coordination of groups of mobile autonomous agents using nearest neighbor rules, *IEEE Transactions on Automatic Control* **48**, 6, pp. 988–1001.

Juhász, F. (1991). The asymptotic behaviour of Fiedler's algebraic connectivity for random graphs, *Discrete Mathematics* **96**, pp. 59–63.

Kaneko, K. (1989). Chaotic but regular posi-nega switch among coded attractors by cluster-size variation, *Physical Review Letters* **63**, 3, pp. 219–223.

Karpelevich, F. I. (1951). On the characteristic roots of matrices with nonnegative elements, *Izv. Akad. Nauk. USSR Ser. Mat.* **14**, pp. 361–383, in Russian.

Kellogg, R. B. (1972). On complex eigenvalues of m and p matrices, *Numer. Math.* **19**, pp. 170–175.

Kolb, B. and Whishaw, I. Q. (1990). *Fundamentals of Human Neuropsychology*, 3rd ed. (W. H. Freeman and Company, New York).

Kumar, R., Raghavan, P., Rajagopalan, S., Sivakumar, D., Tomkins, A. and Upfal, E. (2000). The Web as a graph, in *Proceedings of the nineteenth ACM SIGMOD-SIGACT-SIGART symposium on Principles of database systems*, pp. 1–10.

Lafferriere, G., Williams, A., Caughman, J. and Veerman, J. J. P. (2005). Decentralized control of vehicle formations, *Systems & Control Letters* **54**, 9, pp. 899–910.

Lakshmikantham, V. and Liu, X. Z. (1993). *Stability Analysis In Terms of Two Measures* (World Scientific, Singapore).

Lévy, L. (1881). Sur la possibilité du l'équilibre électrique, *C. R. Acad. Sci. Paris* **93**, pp. 706–708.

Lin, W. and Zhan, X. (2006). Polarized networks, diameter, and synchronizability of networks, Tech. rep., arXiv:cond-mat/0604295.

Lin, Z., Francis, B. and Maggiore, M. (2005). Necessary and sufficient graphical conditions for formation control of unicycles, *IEEE Transactions on Automatic Control* **50**, 1, pp. 121–127.

Lubachevsky, B. and Mitra, D. (1986). A chaotic asynchronous algorithm for computing the fixed point of a nonnegative matrix of unit spectral radius, *Journal of the Association for Computing Machinery* **33**, 1, pp. 130–150.

Lubotzky, A., Phillips, R. and Sarnak, P. (1988). Ramanujan graphs, *Combinatorica* **8**, 3, pp. 261–277.

Milgram, S. (1967). The small-world problem, *Psychology Today* **1**, pp. 61–67.

Miller, S. J., Novikoff, T. and Sabelli, A. (2006). The distribution of the second largest eigenvalue in families of Ramanujan graphs, Tech. rep., math.CO/0611649.

Minc, H. (1988). *Nonnegative Matrices* (John Wiley & Sons, New York).

Mohar, B. (1989). Isoperimetric numbers of graphs, *Journal of Combinatorial Theory, Series B* **47**, pp. 274–291.

Mohar, B. (1991a). Eigenvalues, diameter, and mean distance in graphs, *Graphs and Combinatorics* **7**, pp. 53–64.

Mohar, B. (1991b). The Laplacian spectrum of graphs, in Y. Alavi, G. Chartrand, O. R. Oellermann and A. J. Schwenk (eds.), *Graph Theory, Combinatorics, and Applications*, vol. 2 (Wiley), pp. 871–898.

Mohar, B. (1997). Some applications of Laplace eigenvalues of graphs, in G. Hahn and G. Sabidussi (eds.), *Graph Symmetry: Algebraic Methods and Applications* (Kluwer), pp. 225–275.

Moreau, L. (2003). Leaderless coordination via bidirectional and unidirectional time-dependent communication, .

Moreau, L. (2005). Stability of multiagent systems with time-dependent communication links, *IEEE Transactions on Automatic Control* **50**, 2, pp. 169–182.

Murthy, M. R. (2003). Ramanujan graphs, *Journal of the Ramanujan Mathematical Society* **18**, 1, pp. 1–20.

Nelson, S. and Neumann, M. (1987). Generalization of the projection method with applications to SOR method for Hermitian positive semidefinite linear systems, *Numerische Mathematik* **51**, 2, pp. 123–141.

Nesterov, Y. and Nemirovskii, A. (1994). *Interior-Point Polynomial Algorithms in Convex Programming*, *SIAM Studies in Applied Mathematics*, vol. 13 (Society for Industrial and Applied Mathematics).

Neumann, M. and Schneider, H. (1999). The convergence of general products of matrices and the weak ergodicity of Markov chains, *Linear Algebra and its Applications* , pp. 307–314.

Newman, M. E. J. (2003). The structure and function of complex networks, *SIAM Review* **45**, 2, pp. 167–256.

Newman, M. E. J. and Watts, D. J. (1999). Scaling and percolation in the small-world network model, *Physical Review E* **60**, 6, pp. 7332–7342.

Nishikawa, T., Motter, A. E., Lai, Y.-C. and Hoppensteadt, F. C. (2003). Heterogeneity in oscillator networks: Are smaller worlds easier to synchronize? *Physical Review Letters* **91**, 1, pp. 014101–1–4.

Olfati-Saber, R. and Murray, R. M. (2004). Consensus problems in networks of agents with switching topology and time-delays, *IEEE Transactions on Automatic Control* **49**, 9, pp. 1520–1533.

Paz, A. (1965). Definite and quasidefinite sets of stochastic matrices, *Proceedings of the American Mathematical Society* **16**, pp. 634–641.

Paz, A. (1970). Ergodic theorems for infinite probabilistic tables, *Annals of Mathematical Statistics* **41**, 2, pp. 539–550.

Paz, A. and Reichaw, M. (1967). Ergodic theorems for sequences of infinite stochastic matrices, *Proc. Cambridge Philos. Soc.* **63**, pp. 777–784.

Pecora, L. M. and Carroll, T. L. (1990). Synchronization in chaotic systems, *Physical Review Letters* **64**, 8, pp. 821–824.

Pecora, L. M. and Carroll, T. L. (1998a). Master stability functions for synchronized chaos in arrays of oscillators, in *Proceedings of the 1998 IEEE Int. Symp. Circ. Syst.*, vol. 4 (IEEE), pp. IV–562–567.

Pecora, L. M. and Carroll, T. L. (1998b). Master stability functions for synchronized coupled systems, *Physical Review Letters* **80**, 10, pp. 2109–2112.

Penrose, M. (2003). *Random Geometric Graphs* (Oxford University Press).

Rapoport, A. (1957). Contribution to the theory of random and biased nets, *Bulletin of Mathematical Biology* **19**, 4, pp. 257–277.

Ren, W. and Beard, R. W. (2005). Consensus seeking in multiagent systems under dynamically changing interaction topologies, *IEEE Transactions on Automatic Control* **50**, 5, pp. 655–661.

Ren, W., Beard, R. W. and McLain, T. W. (2004). Coordination variables and consensus building in multiple vehicle systems, in V. Kumar, N. E. Leonard and A. S. Morse (eds.), *Cooperative Control, Lecture Notes in Control and Information Sciences*, vol. 309 (Springer-Verlag), pp. 171–188.

Ren, W. and Stepanyan, V. (2003). Information consensus in distributed multiple vehicle coordinated control, in *Proceedings of the 42nd IEEE Conference on Decision and Control*, pp. 2029–2034.

Seneta, E. (1973). On the historical development of the theory of finite inhomogeneous Markov chains, *Proceedings of the Cambridge Philosophical Society* **74**, pp. 507–513.

Seneta, E. (1979). Coefficients of ergodicity: structure and applications, *Advances in Applied Probability* **11**, pp. 576–590.

Shen, J. (2000). A geometric approach to ergodic non-homogeneous Markov chains, in *Wavelet Analysis and Multiresolution Methods, Lecture Notes in Pure and Applied Math.*, vol. 212 (Marcel Dekker), pp. 341–366.

Sierksma, G. and Hoogeveen, H. (1991). Seven criteria for integer sequences being graphic, *Journal of Graph Theory* **15**, 2, pp. 223–231.

Su, Y. and Bhaya, A. (2001). Convergence of pseudocontractions and applications to two-stage and asynchronous multisplitting for singular m-matrices, *SIAM journal of matrix analysis and its applications* **22**, 3, pp. 948–964.

Swamy, M. N. S. and Thulasiraman, K. (1981). *Graphs, Networks, and Algorithms* (John Wiley & Sons).

Tan, C.-P. (1982). A functional form for a particular coefficient of ergodicity, *Journal of Applied Probability* **19**, pp. 858–863.

Tarjan, R. E. (1972). Depth-first search and linear graph algorithms, *SIAM Journal on Computing* **1**, 2, pp. 146–160.

Taussky, O. (1949). A recurring theorem on determinants, *American Mathematical Monthly* **10**, pp. 672–676.

Vandenberghe, L. and Boyd, S. (1996). Semidefinite programming, *SIAM Review* **38**, 1, pp. 49–95.

Vicsek, T., Czirók, A., Ben-Jacob, E., Cohen, I. and Shochet, O. (1995). Novel type of phase transition in a system of self-driven particles, *Physical Review Letters* **75**, 6, pp. 1226–1229.

Wang, X. F. and Chen, G. (2002). Synchronization in small-world dynamical networks, *International Journal of Bifurcation and Chaos* **12**, 1, pp. 187–192.

Watts, D. J. and Strogatz, S. H. (1998). Collective dynamics of 'small-world' networks, *Nature* **393**, pp. 440–442.

Wiggins, S. (1990). *Introduction to Applied Nonlinear Dynamical Systems and Chaos, Texts in Applied Mathematics*, vol. 2 (Springer-Verlag).

Winkler, R. L. (1968). The consensus of subjective probability distributions, *Management Science* **15**, 2, pp. B61–B75.

Wolfowitz, J. (1963). Products of indecomposable, aperiodic, stochastic matrices, *Proc. Amer. Math. Soc.* **14**, pp. 733–737.

Wu, C. W. (1998a). Global synchronization in coupled map lattices, in *Proceedings of the 1998 IEEE Int. Symp. Circ. Syst.*, vol. 3 (IEEE), pp. III–302–305.

Wu, C. W. (1998b). Synchronization in arrays of chaotic circuits coupled via hypergraphs: static and dynamic coupling, in *Proceedings of the 1998 IEEE Int. Symp. Circ. Syst.*, vol. 3 (IEEE), pp. III–287–290.

Wu, C. W. (2001). Synchronization in arrays of coupled nonlinear systems: passivity, circle criterion and observer design, in *IEEE International Symposium on Circuits and Systems*, pp. III–692–695.

Wu, C. W. (2002). *Synchronization in coupled chaotic circuits and systems* (World Scientific).

Wu, C. W. (2003a). Perturbation of coupling matrices and its effect on the synchronizability in arrays of coupled chaotic circuits, *Physics Letters A* **319**, pp. 495–503.

Wu, C. W. (2003b). Synchronization in coupled arrays of chaotic oscillators with nonreciprocal coupling, *IEEE Transactions on Circuits and Systems–I: Fundamental Theory and Applications* **50**, 2, pp. 294–297.

Wu, C. W. (2005a). Agreement and consensus problems in groups of autonomous agents with linear dynamics, in *Proceedings of 2005 IEEE International Symposium on Circuits and Systems*, pp. 292–295.

Wu, C. W. (2005b). Algebraic connectivity of directed graphs, *Linear and Multilinear Algebra* **53**, 3, pp. 203–223.

Wu, C. W. (2005c). On bounds of extremal eigenvalues of irreducible and m-reducible matrices, *Linear Algebra and Its Applications* **402**, pp. 29–45.

Wu, C. W. (2005d). On Rayleigh-Ritz ratios of a generalized Laplacian matrix of directed graphs, *Linear Algebra and its Applications* **402**, pp. 207–227.

Wu, C. W. (2005e). Synchronizability of networks of chaotic systems coupled via a graph with a prescribed degree sequence, *Physics Letters A* **346**, 4, pp. 281–287.

Wu, C. W. (2005f). Synchronization in arrays of coupled nonlinear systems with delay and nonreciprocal time-varying coupling, *IEEE Transactions on Circuits and Systems–II* **53**, 5, pp. 282–286.

Wu, C. W. (2005g). Synchronization in networks of nonlinear dynamical systems coupled via a directed graph, *Nonlinearity* **18**, pp. 1057–1064.

Wu, C. W. (2006a). On a matrix inequality and its application to the synchronization in coupled chaotic systems, in *Complex Computing-Networks: Brain-like and Wave-oriented Electrodynamic Algorithms, Springer Proceedings in Physics*, vol. 104, pp. 279–288.

Wu, C. W. (2006b). On some properties of contracting matrices, Tech. Rep. arxiv:math.DS/0604457.

Wu, C. W. (2006c). Synchronization and convergence of linear dynamics in random directed networks, *IEEE Transactions on Automatic Control* **51**, 7, pp. 1207–1210.

Wu, C. W. and Chua, L. O. (1994). A unified framework for synchronization and control of dynamical systems, *International Journal of Bifurcation and Chaos* **4**, 4, pp. 979–998.

Wu, C. W. and Chua, L. O. (1995). Synchronization in an array of linearly coupled dynamical systems, *IEEE Transactions on Circuits and Systems–I: Fundamental Theory and Applications* **42**, 8, pp. 430–447.

Xiao, L., Boyd, S. and Lall, S. (2005). A scheme for robust distributed sensor fusion based on average consensus, in *IPSN '05: Proceedings of the 4th international symposium on Information processing in sensor networks* (IEEE Press, Piscataway, NJ, USA), ISBN 0-7803-9202-7, p. 9.

Index